Climate Economics

to Irena Sendler

Climate Economics

Economic Analysis of Climate, Climate Change and Climate Policy

Richard S. J. Tol

Professor of Economics, University of Sussex, UK; Professor of the Economics of Climate Change, Vrije Universiteit Amsterdam, The Netherlands

Edward Elgar
Cheltenham, UK • Northampton, MA, USA

Published by
Edward Elgar Publishing Limited
The Lypiatts
15 Lansdown Road
Cheltenham
Glos GL50 2JA
UK

Edward Elgar Publishing, Inc.
William Pratt House
9 Dewey Court
Northampton
Massachusetts 01060
USA

A catalogue record for this book
is available from the British Library

Library of Congress Control Number: 2014938768

ISBN 978 1 78254 591 0   (cased)
ISBN 978 1 78254 592 7   (paperback)
ISBN 978 1 78254 593 4   (eBook)

Typeset by Servis Filmsetting Ltd, Stockport, Cheshire
Printed and bound in Great Britain by T.J. International Ltd, Padstow

Contents in brief

Full contents

Preface

This book has been over 10 years in the making. I started teaching the economics of climate change in Hamburg in 2001, and have been hampered by the lack of a good textbook ever since. My biggest thanks are therefore to the students in Hamburg, Amsterdam and Sussex who suffered through my attempts to master the material that lies in front of you.

My thoughts on the economics of climate change and climate policy have benefitted from discussions with and papers by many people. I name a few: David Anthoff, Doug Arent, David Bradford†, Ian Burton, Carlo Carraro, Bill Cline, Hadi Dowlatabadi, Tom Downing, Sam Fankhauser, Jan Feenstra†, Brian Fisher, Reyer Gerlagh, Christian Gollier, Paul Gorecki, Cameron Hepburn, Huib Jansen†, Klaus Keller, Sean Lyons, David Maddison, Alan Manne†, Rob Mendelsohn, Bill Nordhaus, Steve Pacala, David Pearce†, Roger Pielke Jr, Katrin Rehdanz, Rich Richels, Roberto Roson, Tom Rutherford, Tom Schelling, Steve Schneider†, Joel Smith, Ferenc Toth, Harmen Verbruggen, Marty Weitzman, John Weyant, and Gary Yohe.

A number of people made useful comments on previous drafts, including David Anthoff, Valentina Bosetti, Mike Mastandrea, Lance Wallace, Bob Ward, Tim Worstall, and three anonymous referees. David Anthoff inspired Chapter 13, and wrote the first draft of its text. The team at Edward Elgar is fantastic.

It is common to devote books about climate change to one's children or grandchildren. I don't see why. My parents told me to think for myself, to work hard, and to get an education. I try to pass this on to my kids, and I'm sure they'll be fine, if I succeed, regardless of what the climate throws at them.

Instead, as a warning against the hubris that pervades climate research and policy, I dedicate this book to the memory of Irena Sendler.

Introduction

This is a textbook on the economics of climate, climate change, and climate policy. The book is structured as follows. Chapter 1 reviews the science of climate change. Chapter 2 discusses sources of and scenarios for greenhouse gas emissions, and options for emission reduction. Chapter 3 turns to the costs of emission reduction, and Chapter 4 to policy instruments for emission reduction. Chapter 5 is an interlude on economic valuation of goods and services not traded on markets. Chapter 6 treats the economic impact of climate change. Chapter 7 discusses the relationship between climate (change) and development. Chapter 8 is on optimal climate policy. Chapter 9 discusses the effect of aggregation (over time, over people, and over possible states of the world) on optimal climate policy. Chapter 10 revisits optimal climate policy again, now in the context of uncertainty and learning. Chapter 11 discusses non-cooperative climate policy. Chapter 12 is on adaptation and adaptation policy.

Chapter 14 is an overview and summary. It provides a basis for a single, one-hour lecture on the economics of climate change and gives a taste of the controversies around climate policy.

Every chapter starts with its key messages. These come in the form of tweets. Accuracy is sacrificed for brevity. I find that tweeting my core message before a class or lecture helps me to focus on what I want and need to say. There is an online quiz for each chapter, designed for revising the material covered, again with a focus on the core messages. Both tweets and quizzes help the students distinguish the forest from the trees.

The quizzes, additional figures, and other supporting material can be found at https://sites.google.com/site/climateconomics/.

Chapters end with suggestions for further reading and exercises. The exercises are designed to expand on the text. There are three sets. First, there are "classical" exercises as "calculate this" and "why would that be?" Second, there are reading assignments for presentation and discussion. Third, there is a set of instructions to build an integrated assessment model and use it to shed light on climate policy. This set of exercises is gathered in Chapter 13.

Which set of exercises (if any) to use depends on the structure and aims of modules and courses.

The material is presented at four levels. Prerequisite material is marked with one star*. This should have been covered in an earlier module. It is here presented for completeness and to refresh readers' memories. Basic material is marked with two stars**. This is suited for a course at bachelor's level. Advanced material is marked with three stars***. This is suited for a course at master's level. Specialist material is marked with four stars****. This is suited for a course at PhD level. In every chapter, there is a reading exercise (for each of the three levels) and suggestions for further reading. The listed papers together form a reader at PhD level.

1

The science of climate change

TWEET BOOK

- The 3 most important anthropogenic greenhouse gases, ambient CO_2, CH_4 and N_2O have risen since the Industrial Revolution. #climateeconomics
- Global mean surface air temperature and global mean sea level rise have gone up too, and snow pack down. #climateeconomics
- Greenhouse gases are transparent to visible light from the sun, but opaque to infrared radiation from Earth. #climateeconomics
- With greenhouse gases in the atmosphere, it is easier for energy to enter the planet than to leave it. #climateeconomics
- Higher greenhouse gas concentrations imply warming – but how much is uncertain as there are many, complex feedbacks. #climateeconomics
- Human CO_2 emissions are a tiny fraction of natural emissions, but natural emissions are balanced by natural uptake. #climateeconomics
- By 2100, the global mean temperature is probably 1–4 degrees Celsius higher than now, depending on scenario and model. #climateeconomics
- Warming will be more pronounced towards the poles, in winter, at night, and over land. #climateeconomics
- Water expands as it warms, and sea levels rise. Land ice melts. By 2100, the sea will probably rise by 0.2–0.6 metres. #climateeconomics
- Some places and times will see more rain, other places and times less. Downpours may well become heavier. #climateeconomics
- Tropical storms will probably not extend their range or increase their frequency. Storms everywhere will intensify. #climateeconomics
- As more CO_2 dissolves in water, oceans will become more acidic. #climateeconomics

1.1   Processes**

Figure 1.1 shows observations of the atmospheric concentration of the three main anthropogenic greenhouse gases – carbon dioxide (CO_2), methane

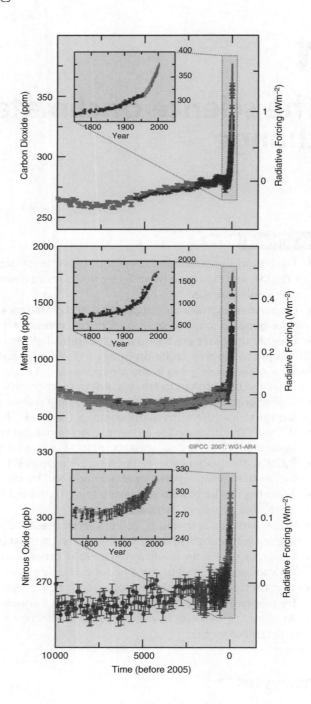

Source: IPCC WG1 AR4 SPM.

Figure 1.1 Atmospheric concentrations of the three main anthropogenic greenhouse gases

(CH_4) and nitrous oxide (N_2O) – over two periods: From the start of the Industrial Revolution (say, 1850) to today, and from the start of the agricultural revolution (say, 8000 BC) to today. Since the start of the Industrial Revolution, ambient greenhouse gases have been on the rise. The increase in the last 150 years is quite unusual given the experience of the last 12 000 years.

Measuring the composition of the atmosphere is a recently developed skill. Older measurements are obtained as follows. As snow falls on ice caps, little bubbles of air are trapped in the newly formed ice. Older air can be found in older ice, deeper in the ice cap. The atmospheric concentration of ancient times can be reconstructed from cores drilled from the ice. Such reconstructions are imperfect, both with regard to the timing and the assumption that air bubbles are hermetically sealed.

Figure 1.2 shows observations of the global mean surface air temperature, the global mean sea level, and the average snow cover in the northern hemisphere. Temperature and sea level have gone up over the last 250 years, and snow cover has declined. This is exactly as one would expect if greenhouse gas concentrations are rising (although climate could also have changed for other reasons).

Figure 1.3 illustrates why. The sun sends energy into space in every direction. Some of that energy is in the part of the spectrum that is visible to the human eye, and some of that energy reaches Planet Earth. The planet is in energy balance: It receives as much energy as it emits, at least on average. If not, the planet would forever heat or cool. Earth does not emit visible light – it is dark at night – but it does emit infrared radiation. Greenhouse gas molecules are, by definition, transparent to visible light but intransparent to infrared radiation. That is, solar energy passes unhindered through the atmosphere, but infrared radiation is absorbed by the greenhouse gas molecules. These molecules get excited, but later return to their base state, emitting energy in any direction. That is the crucial part of the greenhouse effect. Infrared radiation from Planet Earth is directed towards outer space. Infrared radiation from greenhouse gas molecules can go anywhere, including back to the surface. Therefore, if there are greenhouse gases in the atmosphere, it is harder for energy to leave the planet than if there are no such gases. The planet is still in energy balance – incoming energy equals outgoing energy – but more energy is stored on the planet: It is warmer.

The greenhouse effect was first described by Fourier in 1827. The details were worked out by John Tyndall in the 1860s. In 1896, Svante Arrhenius

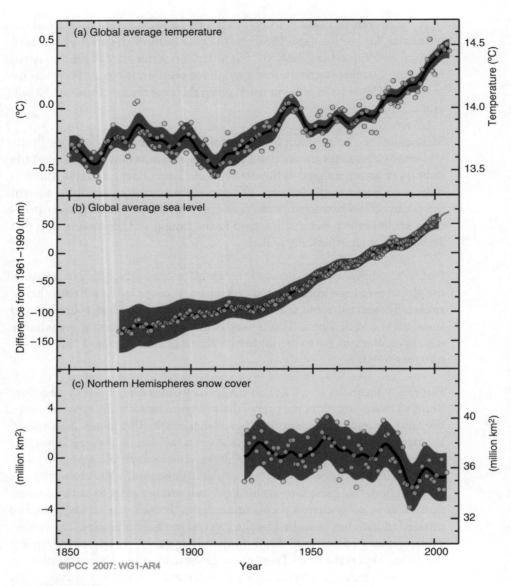

Source: IPCC WG1 AR4 SPM.

Figure 1.2 Observed global mean temperature, global mean sea level, and northern hemispheric snow pack

reckoned that the burning of fossil fuels would increase the concentration of carbon dioxide in the atmosphere, and that this would enhance the greenhouse effect and warm the planet. Figure 1.1 and 1.2 show that this is indeed the case – at least, qualitatively.

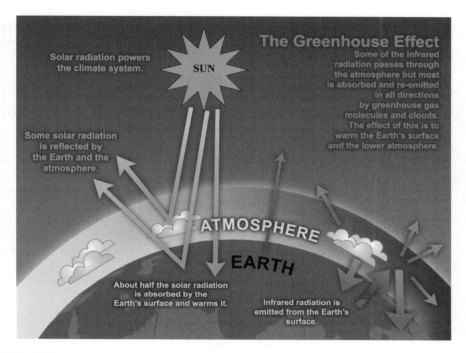

Source: IPCC WG1 AR4 Ch1.

Figure 1.3 The greenhouse effect

Figure 1.4 illustrates some of the complications. It shows radiative forcing, the change in energy per square metre, since 1850. Carbon dioxide is by far the most important substance in the change in the Earth's energy balance. It is also relatively well-known, the main uncertainty being the atmospheric concentration in pre-industrial times. Put together, the other anthropogenic greenhouse gases have contributed about two-thirds as much as carbon dioxide to the total radiative forcing. Relative uncertainty is about as large.

But the human interference with the climate system does not end there. Ozone is a greenhouse gas too. It is not emitted by human activities, but results from interactions in the atmosphere with substances that are emitted by humans. Near the surface, ozone concentrations are higher than they used to be because of precursor emissions from transport and agriculture. Higher up in the atmosphere, ozone concentrations are lower because of emissions of chlorofluorocarbons.

Water vapour is a greenhouse gas too, in fact the most important of them all, but its concentration is only marginally affected by human activity: The

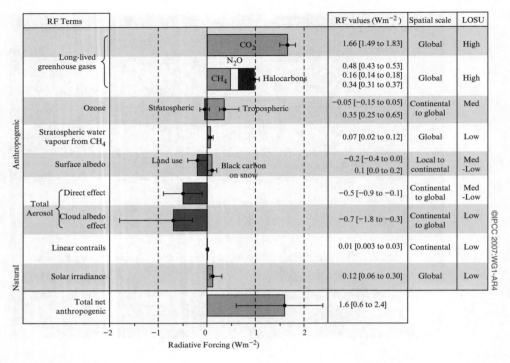

RF Terms		RF values (Wm^{-2})	Spatial scale	LOSU
Long-lived greenhouse gases	CO$_2$	1.66 [1.49 to 1.83]	Global	High
	N$_2$O	0.48 [0.43 to 0.53]		
	CH$_4$ Halocarbons	0.16 [0.14 to 0.18] 0.34 [0.31 to 0.37]	Global	High
Ozone	Stratospheric Tropospheric	−0.05 [−0.15 to 0.05] 0.35 [0.25 to 0.65]	Continental to global	Med
Stratospheric water vapour from CH$_4$		0.07 [0.02 to 0.12]	Global	Low
Surface albedo	Land use Black carbon on snow	−0.2 [−0.4 to 0.0] 0.1 [0.0 to 0.2]	Local to continental	Med -Low
Total Aerosol Direct effect		−0.5 [−0.9 to −0.1]	Continental to global	Med -Low
Cloud albedo effect		−0.7 [−1.8 to −0.3]	Continental to global	Low
Linear contrails		0.01 [0.003 to 0.03]	Continental	Low
Solar irradiance		0.12 [0.06 to 0.30]	Global	Low
Total net anthropogenic		1.6 [0.6 to 2.4]		

Radiative Forcing (Wm^{-2})

©IPCC 2007:WG1-AR4

Figure 1.4 Radiative forcing and its components since pre-industrial times

breakdown of methane (CH$_4$) in the atmosphere increases the concentration of water vapour.

Humans have also changed the albedo, which determines the amount of energy reflected by the surface of Planet Earth. Soot has made snow and ice darker than they used to be, thus absorbing more energy. Less snow and ice also means a darker surface. On the other hand, relatively dark trees have been replaced by relatively light grass.

Particles in the atmosphere from fossil fuel combustion also change the radiative balance, by directly reflecting sunlight and by assisting in the formation of clouds, which also block radiation. The water vapour from aircraft also forms clouds – contrails – but their contribution to global warming is minimal.

Besides the human influences on the climate, there are natural effects as well. Volcanic eruptions can have a rather large, but typically short-lived impact. There is no reason to believe that there is a long-term trend in volcanic activity. There is a trend in the energy output of the sun, but this is small compared with the changes in greenhouse gas concentrations. The slow dynamics of the deep ocean induce semi-regular cycles in the atmosphere with characteristic life-times of years and decades, maybe longer.

Our confidence in the radiative forcing of greenhouse gas emissions is higher than in other radiative forcing, partly because the physics and chemistry of the relevant process is not completely understood (as it is much more complex than the greenhouse effect) and partly because data for pre-industrial times are spotty.

The uncertainty about climate change is much larger than the uncertainty about radiative forcing. The degree of global warming is determined by the amount of radiative forcing and a number of feedbacks in the climate system. Most importantly, warmer air contains more water vapour, and water vapour is a greenhouse gas. The big uncertainty is about cloud formation. Clouds can keep the heat of the sun out, but also the warmth of the earth in. The physics of cloud formation is rather complex and operates at a spatial scale much finer than can be resolved by climate models. Different models therefore use different cloud parameterizations, which behave roughly the same in the current climate but differently in altered climates.

The oceans are another major uncertainty. If the atmosphere warms, so do the waters at the ocean surface. If surface waters warms, so do the waters deeper down. The speed at which energy dissipates into the ocean determines the speed at which the atmosphere warms. The rate of ocean warming depends on a complex pattern of horizontal and vertical currents. Observations of the deep ocean are few and recent, so ocean circulation models are poorly constrained by data.

There are other complications. Fossil fuel combustion also emits aerosols, which directly affect radiation passing through the atmosphere and play a role in cloud formation (and thus indirectly affect the radiative balance). Humans have changed the landscape at a massive scale, and hence the colour and thus albedo of the surface. And humans affect the nutrient cycles (nitrogen, phosphate) and so vegetation, albedo and carbon cycle.

Model uncertainty is reflected in Figure 1.5. It shows – globally, over land, over ocean, and for the six inhabited continents – the observed mean surface air temperature, smoothed over time, for the 20th century. Figure 1.5 also shows the range of model reconstructions for two sets of scenarios: One with all known radiative forcing, and one with only natural forcing. If all forcing

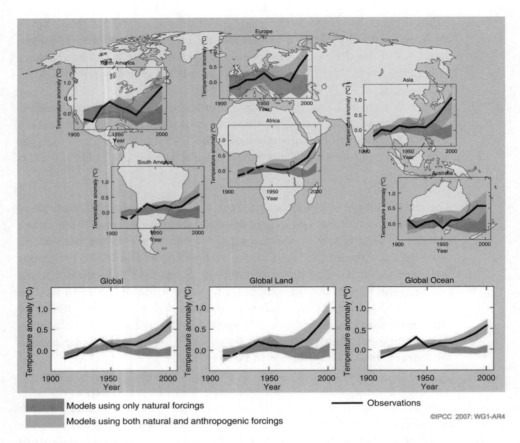

Models using only natural forcings — Observations
Models using both natural and anthropogenic forcings

©IPCC 2007: WG1-AR4

Source: IPCC WG1 AR4 SPM.

Figure 1.5 Observed and modelled mean surface air temperatures: world, land, ocean, continents

is included, the observed warming is somewhere in the model of the predicted range of warming. If anthropogenic forcing is omitted, the observed warming is outside the predicted range. This indicates that it is unlikely, but not impossible, that the observed warming is not, at least partially, to blame on human activity.

Figure 1.6 depicts the carbon cycle, relating the stocks and flows of carbon dioxide. In pre-industrial times, the main exchanges of CO_2 were between the atmosphere, the ocean, and terrestrial vegetation. Each stores a large amount of CO_2. CO_2 fluxes are large too, as vegetation grows in summer and dies back in winter. There is another large stock of carbon in fossil fuels. In natural circumstances, this stock does not play a significant part in

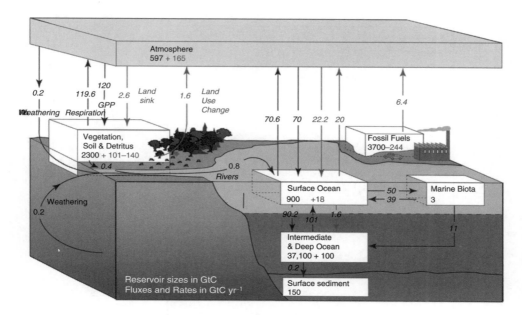

Figure 1.6 depiction of the carbon cycle with the following labels and values:

Atmosphere
597 + 165

0.2
119.6
120
GPP
Weathering Respiration

2.6 Land sink
1.6 Land Use Change
6.4

70.6 70 22.2 20

Vegetation, Soil & Detritus
2300 + 101–140

Fossil Fuels
3700–244

0.4
0.8
Rivers

Surface Ocean
900   +18

50 →
← 39

Marine Biota
3

Weathering
0.2

90.2 101 1.6

11

Intermediate & Deep Ocean
37,100 + 100

0.2

Reservoir sizes in GtC
Fluxes and Rates in GtC yr⁻¹

Surface sediment
150

Notes:
Stocks are in boxes, flows in arrows. Black numbers denote pre-industrial values, red numbers changes since.
Source: IPCC WG1 AR4 Ch7.

Figure 1.6 The carbon cycle

the carbon cycle. However, human exploitation has mobilized this carbon. Emissions of CO_2 from fossil fuel combustion are small compared with natural emissions – but unlike natural emissions, there is no counterbalancing flux. Although human emissions are partly absorbed in the vegetation and ocean, the atmospheric concentration of CO_2 has increased, enhancing the greenhouse effect.

1.2   Projections**

Figure 1.7 shows model reconstructions of the 20th century. Figure 1.7 applies the same models to the 21st century for a range of emission scenarios[1] (see Chapter 2). Over the 21st century, global warming will probably be between 1.4 and 4.2°C, on top of the 0.8°C warming in the 20th century.

Figure 1.8 shows the spatial pattern of warming. Warming is more pronounced over land than over water, and towards the poles. Figure 1.8, warming is more pronounced in winter than in summer, and at night than at day. Models agree on these broad patterns.

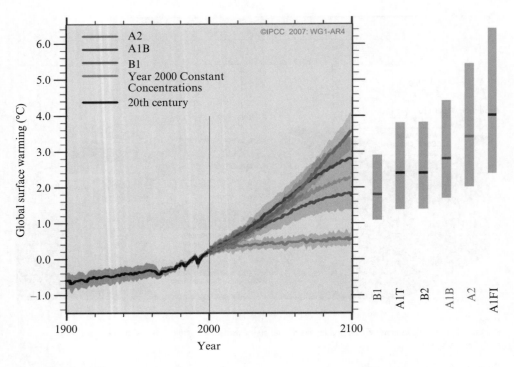

Source: IPCC WG1 AR4 SPM.

Figure 1.7 The global mean surface air temperature as observed for the 20th century and as projected for the 21st century

There is less agreement on the pattern of changes in rainfall. See Online Figure 1.1. On large parts of the global, models do not even agree on the sign of change. However, (sub)tropical areas are likely to get drier and higher latitude areas wetter – this implies that, on the southern hemisphere, more rain will fall over sea (which is no use). In temperate areas, winters will get wetter and summers drier. As it gets warmer, rainfall will tend to get more intense, with heavier downpours in between longer dry spells.

Storms, both in the tropics and elsewhere, are likely to become more intense too. Maximum wind speeds will probably increase. There is no reason to assume that the frequency of storms will change much; or that tropical storms will extend their area.

Water expands if it gets warmer. Sea levels will therefore rise. This is a surprise to some people. After all, tea does not visibly shrink as it cools down. However, the ocean is on average three kilometres deep. If ocean water expands by 0.01%, then sea levels rise by 30 cm. The projected sea level

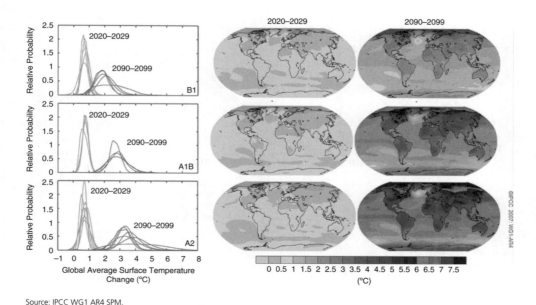

Source: IPCC WG1 AR4 SPM.

Figure 1.8 The spatial pattern of projected warming

rise over the 21st century due to thermal expansion is somewhere between 10 and 40 cm. See Figure 1.9, which also illustrates the jumble of model results that is characteristic of all things climate. The melting of small ice caps and glaciers will add another 10 cm or so to sea level rise. Although glaciers are impressive to the human eye – and their disappearance dramatic – they contain little water relative to the oceans. The melting of floating ice, common around the North Pole, does not contribute to sea level rise, because that ice already displaces sea water. The large ice caps and shelves on Greenland and Antarctica rest on land and do contain a substantial amount of water. If the West-Antarctic Ice Shelf would melt or slide into the sea – the latter could happen much more quickly – sea levels would rise by 5–6 metres. If the Greenland Ice Cap would melt, sea levels would rise by 6–7 metres. If the ice on East-Antarctica would melt, sea levels would rise by some 60 metres. The ice on West-Antarctica and Greenland may not survive the current millennium but will make it to the end of the century.

Sea level rise is not spatially uniform as shown in Online Figure 1.2. This is because warming is not uniform and water is transported by ocean currents. The volume of ice in Antarctica is such that gravity pulls water to the South Pole. Should that ice melt, sea levels would on average rise by 70 metres or so. Sea level rise in Europe would be some 100 metres as the water is more evenly distributed over the globe.

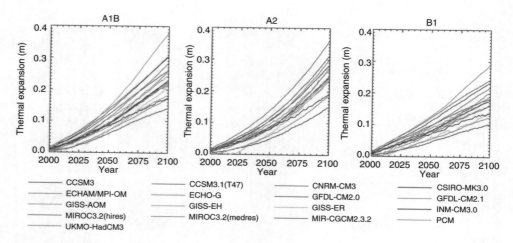

Source: IPCC WG1 AR4 Ch10.

Figure 1.9 Projected sea level rise for the 21st century

Figure 1.6 shows that there is a lot of CO_2 in the ocean. Marine biota contains only a relatively small amount of carbon. The bulk of the carbon is dissolved in water. The partial pressure of CO_2 in the atmosphere equals the partial pressure of CO_2 in the ocean. Thus if there is more CO_2 in the atmosphere, there will be more CO_2 in the ocean. The proper name of carbon dioxide (dissolved in water) is carbonic acid. Higher CO_2 concentrations in ocean waters therefore imply a more acidic ocean, or rather a less alkaline one.

FURTHER READING

Every six years, Working Group I of the Intergovernmental Panel on Climate Change publishes a major assessment of the natural science of climate change. The information is layered, with a Summary for Policy Makers with high-level information, Technical Summaries with more detail, and multiple chapters with a lot of detail and references to the underlying literature. These reports can be found at http://www.ipcc.ch/.

Climate research is rather controversial. Good introductions to the controversy are Mike Hulme's book *Why We Disagree About Climate Change: Understanding Controversy, Inaction, and Opportunity* (2009), Donna Laframboise' book *The Delinquent Teenager Who Was Mistaken For The World's Top Climate Expert* (2011) and Andrew W. Montford's book *The Hockey Stick Illusion: Climategate and the Corruption of Science* (2010).

NOTE

1 Successful prediction is the ultimate aim of positive research. Prediction comes in gradations. "The sun will rise tomorrow at 5:28" is an unconditional prediction (and wrong in most places at most days). "The Earth will warm by 3°C if the atmospheric concentration of carbon dioxide doubles" is a conditional prediction; it depends on the change in atmospheric carbon dioxide. If our description of future events is incomplete, as it is in most cases and certainly in climate change, predictions are necessarily conditional. In climate

research, people prefer the terms "scenario" and "projection" over "conditional prediction". Scenarios of future emissions are conditional predictions, in a way, but system boundaries are not well understood – the prediction is conditional on something vague – and relative probabilities are not estimated. Projections of future climate change are predictions conditional on future emissions, and conditional on the initial state of the climate; because this initial state is not known completely and precisely, and because the climate system is chaotic, a projection is a realization from a stochastic process; it is not the mean or the mode or a known percentile. Predictions play a different role in normative research, of course. There is the Lucas Critique: A prediction in a self-aware system will change the system. More practically, predictions are conditional on policy, and often aim to change policy. "If you do not intervene, dear policy maker, bad things will happen." In a policy context, predictions are intended to be self-defeating prophesies. A false prediction is thus a sign of success rather than failure.

2

Emissions scenarios and options for emission reduction

TWEET BOOK

- Fossil fuel combustion is the main source of CO_2. Per unit of energy, coal emits most, followed by oil and gas. #climateeconomics
- Land use change and cement production are other sources of CO_2. Specialized industries emit halocarbons. #climateeconomics
- Methane results from paddy rice, livestock, waste, and gas leakage, nitrous oxide from agricultural soils. #climateeconomics
- CO_2 emissions equal population times income per capita times energy use per output times emissions per unit of energy. #climateeconomics
- World CO_2 emission +2.1%/year 1970–2008: population +1.5%, income +1.5%, energy per GDP –0.9%, CO_2 per energy –0.01%. #climateeconomics
- Existing scenarios of future greenhouse gas emissions do not reflect the full range of historical experience. #climateeconomics
- There's not enough conventional oil and gas to substantially change climate. Future climate is driven by their replacements. #climateeconomics
- Reduced population and economic growth would reduce emissions, but few elected governments would opt for this. #climateeconomics
- Technological change reduces emissions but current effort would need to be trebled to stabilize emissions. #climateeconomics
- Behavioural change reduces emissions too, but habits are hard to change and market imperfections waste a lot of energy. #climateeconomics
- Carbon-free fuels are another option but nuclear and hydro power are unpopular. #climateeconomics
- Renewables are (very) expensive; volatile and unpredictable; and bring their own environmental problems. #climateeconomics

- CO_2 can be captured and stored at a price. Scale, permanence, and safety are issues. CCS is end-of-pipe solution. #climateeconomics
- Slowing deforestation would reduce emissions but if that were easy it would have been done long ago. #climateeconomics
- Geoengineering is a risky option. There are concerns about who would decide to geoengineer the global climate. #climateeconomics

2.1   Sources of greenhouse gas emissions**

There are a number of different greenhouse gases. Figure 1.4 shows their relative contribution since pre-industrial times. Figure 2.1 shows the relative contributions in the year 2000.

Carbon dioxide is the most important anthropogenic greenhouse gas. Fossil fuel combustion is the main source of CO_2. Fossil fuels are carbohydrates. As they are burned, the chemical bond between the carbon and the hydrogen is broken. Both are oxidized, to CO_2 and H_2O, respectively. In this process, net energy is released. CO_2 emissions are thus intrinsic to the process: You cannot get energy out of fossil fuel without forming CO_2.

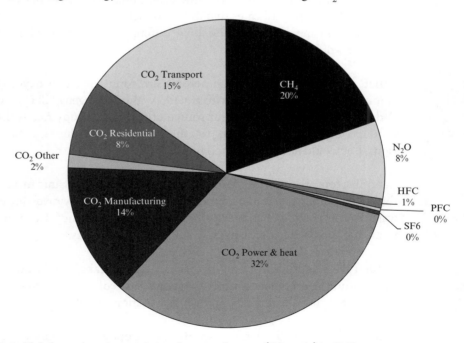

Figure 2.1 Global greenhouse gas emissions by gas and source (CO_2 only) in 2000

Fossil fuels come in a number of varieties. Peat emits most CO_2 per unit of energy (99–117 tCO_2/TJ), followed by coal (98–109 tCO_2/TJ), oil (73–77 tCO_2/TJ), and natural gas (56–58 tCO_2/TJ).

Land use change is another major source of CO_2. Plants are made of carbohydrates too. As tall trees have been replaced by small grass, less carbon is stored in terrestrial vegetation. A substantial part of the wood was burned and CO_2 formed.

Cement production is the least important source. CO_2 is vented as limestone is transformed to cement.

Methane is the next most important greenhouse gas. Ruminants (cows etc.) are a main source. Grass and meat are both carbohydrates, but there are more hydrogen atoms per carbon atom in grass than in meat. Ruminants have therefore formed a symbiotic relationship with methanogenic bacteria, sacrificing one carbon atom to remove four hydrogen atoms. The methane is then burped out. This is an ancient relationship, shared by a large number of grazing animals. Marsupials (kangaroos etc.) use a different solution: acetate rather than methane. Considering the evolutionary distances between cows and kangaroos, this indicates that milk production would be hard to achieve without emitting methane, and how different meat production without methane would be.

When plant material rots in an aerobic environment (with oxygen), CO_2 is formed. In an anaerobic environment (without oxygen), CH_4 is formed. Paddy rice is thus another major source of methane. Paddy rice is the most productive grain crop. Switching to other crops would reduce methane emissions, but would also reduce food production.

Landfills, too, are anaerobic environments with a lot of organic material and thus high methane emissions. Emissions can be reduced by diverting organic waste to composting or incineration; or by capping the landfill, capturing the methane, and flaring it or using it to substitute natural gas.

Natural gas is another word for methane. Methane leaks into the atmosphere from natural gas exploitation and transport. Gas is often found together with coal and oil, and is emitted from their exploitation as well.

Nitrous oxide is the third most important anthropogenic greenhouse gas, primarily emitted from agricultural soils that have been treated with nitrogenous fertilizers. Emission reduction is thus hard without affecting food production.

There are also a range of industrial greenhouse gases. Most of these are artificial: They do not occur naturally. Most were invented after World War II to serve particular purposes – coolants and propellants are two prominent examples. Other gases are by-products of industrial processes – semiconductor manufacturing and packaging material are two important examples. Although the absolute volumes of these emissions are small, these gases tend to be particularly potent greenhouse gases and some have an atmospheric life-time that is measured in tens of thousands of years. Emission reduction is feasible through the development of substitute processes or products, and in select cases through improved waste management.

2.2 Trends in carbon dioxide emissions**

The Kaya identity is a useful tool to understand trends in emissions. If applied to carbon dioxide from fossil fuel combustion, it looks as follows:

$$M = P\frac{YEM}{PYE} \tag{2.1}$$

where M denotes emissions, P population, Y Gross Domestic Product, and E primary energy use. Thus the Kaya identity has that emissions equal the number of people times per capita income times energy intensity (energy use per unit of economic activity) times carbon intensity (emissions per unit of energy use). This is an identity. All but one term on the right-hand side of Equation (2.1) cancel so that $M=M$.

Although an identity, it is useful, and perhaps more so if expressed in proportional growth rates. Take logs on both side of Equation (2.1) and the first partial derivative to time. Then

$$\frac{\partial \ln M}{\partial t} = \frac{\partial \ln P}{\partial t} + \frac{\partial \ln (Y/P)}{\partial t} + \frac{\partial \ln (E/Y)}{\partial t} + \frac{\partial \ln (M/E)}{\partial t} \tag{2.2}$$

Thus[1] the growth rate of emissions equals the growth rate of the population plus the growth rate of per capita income plus the growth rate of energy intensity plus the growth rate of carbon intensity.

Figure 2.2 shows global carbon dioxide emissions between 1970 and 2008. CO_2 emissions rose by 2.1% per year. Why? The Kaya identity allows us to interpret past trends. Population growth was 1.5% per year over the same

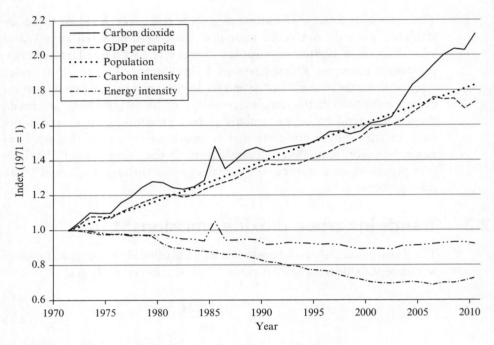

Figure 2.2 Global emissions of carbon dioxide and its constituents

period. Emissions per capita thus rose by 0.6% per year. Per capita income rose by 1.5% per year, again slightly slower than the emissions growth rate. Total income thus rose by 3.0% per year, much faster than emissions. This is primarily because the energy intensity of production fell by 0.9% per year. The carbon intensity of the energy system also fell, but only by 0.01% per year.[2]

The Kaya identity also allows us to project emissions into the future. We need to build a scenario of population growth, economic activity, energy use, and energy supply. See Section 0.

Finally, the Kaya identity allows us to assess how emissions can be cut. We would need to reduce population or income, or improve energy or carbon efficiency. See Section 0.

2.3 Scenarios of future emissions**

Figure 2.3 shows three alternative scenarios of future climate change (plus a diagnostic scenario). These scenarios are based on assumptions about population, economy, and technology. Such assumptions are not independent of

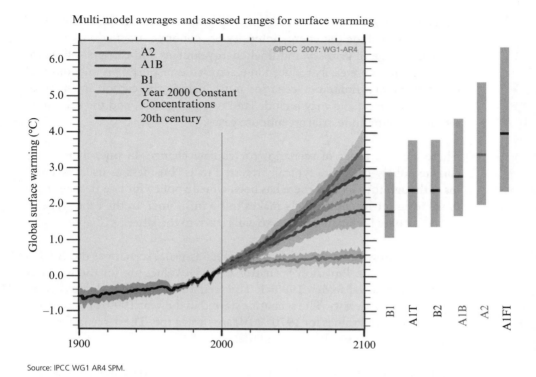

Multi-model averages and assessed ranges for surface warming

Source: IPCC WG1 AR4 SPM.

Figure 2.3 The global mean surface air temperature as observed and as projected

one another. For example, poorer people tend to have shorter lives and more children. Technological progress drives economic growth.

Although our understanding of the processes of long-term development has considerably improved in recent decades, it does not permit any confidence in forecasts over a century or longer. Therefore, scenarios are built instead. Scenarios are not predictions. Scenarios are not-implausible, internally consistent storylines of how the future might unfold.

As stated above, the Kaya identity is useful for organizing future scenarios. Emission scenarios must include the number of people, but may also have their age structure – because that drives decisions on consumption and saving and hence economic growth – their education – because that drives labour productivity and hence growth – and urbanization – because that drives travel and transport and hence energy use. Emission scenarios must include per capita income, but may also have the structure of the economy – because certain sectors use more energy per unit value added than others – and expenditure patterns – because a beef- and rice-based diet emits more methane than a mutton- and wheat-based diet. Emission scenarios must include the energy intensity of economic production, and may include a range of primary and final energy sources and carriers – because emissions are more easily reduced in electrified transport than in liquid-fuel based transport. Emission scenarios must include the carbon intensity of the energy sector, and may include land use, agriculture, and the economic sectors that emit industrial greenhouse gases.

There are two types of scenarios for climate change. In one, there is no climate policy. These are typically referred to as "business-as-usual" scenarios, although the fact that there has been climate policy for two decades now in some countries increasingly makes this a misnomer. In the other type of scenario, there is climate policy. We will return to the latter in Chapter 8.

Figure 2.4 shows a key example of business-as-usual scenarios: the IPCC's Special Report on Emission Scenarios (SRES). Values are for the world as whole, and indexed so that 1990=1. The scenarios are broken down according to the Kaya identity. The scenarios started in the year 1990. For comparison, the observed values for 1970–2010 are shown too. These scenarios were implemented with six alternative models.

Figure 2.4 shows the mean plus or minus twice the standard deviation across these models.

There are three scenarios for population, and four for the rest. This implies that the IPCC assumed that population growth is independent of per capita income, an assumption at odds with everything we know about fertility and mortality. All scenarios of per capita income show exponential growth, and some very rapid growth, even though some parts of the world have enjoyed little growth in the past. All scenarios show a steady improvement of energy efficiency, at a rate that exceeds the experience of the last 40 years. All scenarios show a steady fall in carbon intensity, even though recent history showed both decreases and increases. Although peculiar, the SRES scenarios form the basis of much research on climate change, its impacts, and policies to reduce greenhouse gas emissions.

The availability of fossil fuels is a crucial part of any scenario of future carbon dioxide emissions. Figure 2.5 shows estimates of the reserves and resources of fossil fuels by type. The estimates are taken from the World Energy Council's Survey of Energy Resources of 2010, when the shale gas revolution was tentatively reaching beyond the borders of the USA and the shale oil revolution was in its infancy. Reserves can be profitably exploited

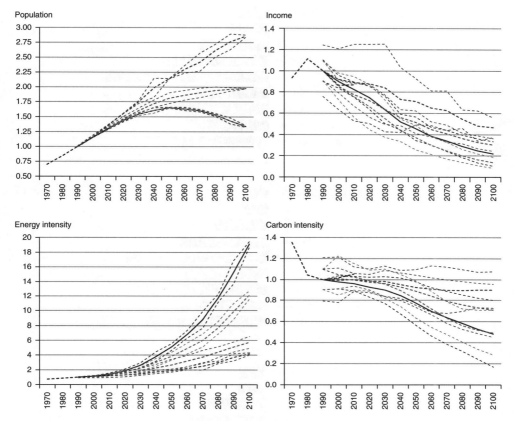

Notes:
A1 brown – low population, very high income; A2 blue – high population, low income; B1 orange – low population, high income; B2 green – middle population, middle income.

Data are indexed to 1990=1; dotted lines are the 95% confidence interval.

Figure 2.4 The SRES scenarios for the world broken down according to the Kaya identity

with current technology at current prices and costs. Resources are known or suspected to be there, and may become commercial in the future. Figure 2.5 reveals that conventional oil and gas reserves are relatively small: 317 billion tonnes of oil equivalent. In 2009, total primary energy use was 11.6 GTOE. There is therefore enough conventional oil and gas to cover energy demand for another 27 years. Figure 2.5 also reveals, however, that there are plenty of other types of fossil fuels, including coal of course but also large resources of unconventional liquids and gases.

The second panel of Figure 2.5 shows the carbon dioxide emissions that would result if these fossil fuels were burned. For comparison, global 2008

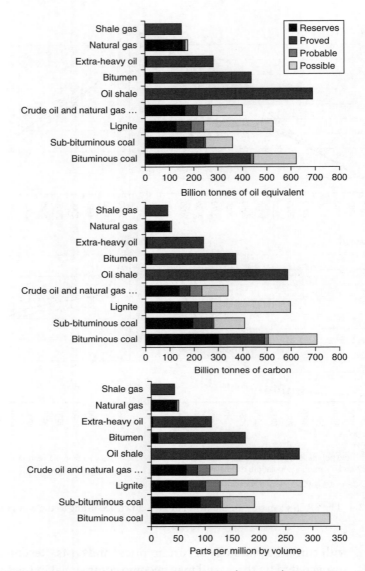

Figure 2.5 Fossil fuel reserves and resources as estimated for 2010 (top panel), their carbon content (middle panel), and implied carbon dioxide concentrations (bottom panel)

emissions were 30 billion tonnes of CO_2. We can keep up current emissions for 100 years or more. The third panel shows the impact on the atmospheric concentration, should all available fossil fuels be burned at once. Conventional oil and gas can contribute only about 100 ppm. Other fossil fuels, reserves and resources, are worth another 1500 ppm.

This implies that the climate problem is not driven by conventional oil and gas, but rather by what will replace conventional oil and gas when they run out. The future energy sector will therefore be very different. Different companies and countries will dominate. Technologies will be different too, and trillions of dollars will be invested in new equipment and infrastructure.

2.4   Options for emission reduction**

The Kaya identity identifies the main options for emission reduction.

Some murderous regimes in Africa actively seek to reduce the population of their countries. Few democratic countries would seek to emulate this in the name of climate policy. Indeed, population policy is controversial in most democracies. China, however, has often put forward its one-child policy as one of its major contributions to climate policy.

The collapse of the former Soviet Union and its aftermath has shown that reducing the level of per capita income is an effective way of cutting greenhouse gas emissions. The Great Recession further demonstrated the power of economic growth over emissions growth. Promoting slower economic growth is not recommended to a politician seeking re-election.

That leaves us with just two of the four terms in the Kaya identity.

Energy efficiency improvements have kept the rise of carbon dioxide emissions in check (see Figure 2.2). Energy efficiency is likely to further improve in the future regardless of climate policy. This is because energy is a cost. A gadget that is the identical to its competitor but uses less energy is more appealing to customers. Companies therefore invest in improving the energy efficiency of their products.

Energy efficiency improvement does not necessarily imply reduced energy use. For instance, the fuel efficiency of the US car fleet was roughly constant between 1980 and 2010. This is a remarkable feat of engineering as, over the same period, the size and weight of cars increased considerably. The gains in fuel efficiency were used not to reduce energy use, but rather to increase comfort. This is known as the rebound effect. Better energy efficiency means lower energy costs means higher energy use. Improving the insulation of homes, for instance, often leads to higher indoor temperatures at the expense of reduced energy use.

There are a number of other options not captured by the Kaya identity. Above, the Kaya identity was interpreted for carbon dioxide emissions from fossil fuel combustion.

The Kaya identity is about the structural causes of emissions and structural solutions. There is also an end-of-pipe solution: Carbon capture and storage (CCS). In CCS, carbon dioxide is separated before, during, or after burning. It is then captured and transported to be stored in a safe place. Carbon capture requires capital and energy. The investment cost of a power plant with capture is some 25% higher than that of a similar plant without, and some 30% of the energy output of the plant will be devoted to carbon capture. Transport is costly too. According to some estimates, if we want to capture all carbon dioxide from power generation, the transport network would be several times bigger than the network for oil and gas. The main issues with storage are permanence and safety. There is little point in storing carbon dioxide if it leaks out again. Sudden releases of carbon dioxide would endanger animal and human life.

Besides the emissions from fossil fuel combustion, land use change also releases carbon dioxide. Reducing such emissions requires slowing down the pace of deforestation, or even reversing it. There are other reasons for doing so. Tropical forests are rich in biodiversity. Forests upstream protect against floods downstream. Mangrove forests shield coasts from waves and wind, and provide food and shelter for animals. Agroforestry promotes soil conservation and crop diversification. Yet, despite many attempts to slow deforestation, it has continued apace. This suggests that it is difficult and expensive. Unless a more lucrative alternative is offered to those that decide to chop down trees, they will continue.

Note that climate policy may even accelerate deforestation. Bioenergy needs land. Palm oil plantations in Southeast Asia replace virgin forest. Sugarcane farms in South America push other crops onto pasture land, and pasture into the rainforest.

As discussed above, methane emissions are intrinsic to the production of dairy, rice, and certain types of meat. Although technical measures can be used to reduce emissions by a little bit, more substantial emission reduction requires volume measures – less dairy, less rice, different meat. Reducing nitrous oxide emissions requires more judicial use of fertilizers and other crop management practices, lest food production fall. Methane from waste disposal and mining can be captured and either flared or burned as a fuel. Almost all emissions can be captured with sufficiently high investment.

Similarly, leaks in gas pipes can be fixed to any standard one is willing to pay for. Industrial gases can be replaced with other substances, which at present are either more expensive or perform worse.

Finally, there is geoengineering. The aim of geoengineering is not to prevent climate change, but rather to change the climate back. There are many ways to achieve this, from spraying water over the oceans to putting aerosols in the atmosphere and mirrors in space. Geoengineering sounds attractive at first sight, as it is cheap and does not require a large number of countries to cooperate. However, uncertainty is one of the main features of climate science. If we do not really know the consequences of putting carbon dioxide in the atmosphere, do we think we know how much sulphur aerosols we should put where to offset the impact of carbon dioxide? Even if successful, geoengineering is risky. With climate change solved, how do you convince policy makers to continue to invest in geoengineering for decades, maybe centuries on end?

FURTHER READING

Every six years, Working Group III of the Intergovernmental Panel on Climate Change publishes a major assessment of options for emission reduction. The information is layered, with a Summary for Policy Makers with high-level information, Technical Summaries with more detail, and multiple chapters with a lot of detail and references to the underlying literature. These reports can be found at: http://www.ipcc.ch/. In 2000, the IPCC released a Special Report on Emissions Scenarios, which can be found at the same site. More recent IPCC scenarios are referred to as *Representative Concentration Scenarios* and *Shard Socio-Economic Pathways* (van Vuuren et al., 2011, *Climatic Change*).

IDEAS/RePEc has a curated bibliography on this topic at http://biblio.repec.org/entry/tda.html.

EXERCISES

2.1 Between 1970 and 2004, how fast should energy efficiency improvements have been to keep global carbon dioxide emissions constant? And carbon efficiency improvements? How do these numbers compare with the observed trend?

2.2 Figure 2.6 shows the Kaya identity for agricultural emissions of methane and nitrous oxide. How would you define the Kaya identity in this case? Discuss the results.

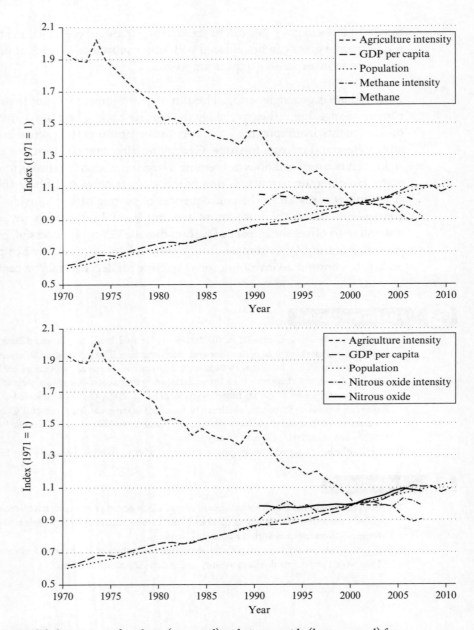

Figure 2.6 Global emissions of methane (top panel) and nitrous oxide (bottom panel) from agriculture and its constituents

2.3 Read and discuss:

- **D. Helm, R. Schmale and J. Philips (2007), *Too Good to be True? The UK's Climate Change Record*. http://www.dieterhelm.co.uk/sites/default/files/Carbon_record_2007.pdf.
- **D. Diakoulaki and M. Mandaraka (2007), 'Decomposition analysis for assessing the progress in decoupling industrial growth from CO_2 emissions in the EU manufacturing sector', *Energy Economics*, **29**, 636–664.
- ***G.P. Peters and E.G. Hertwich (2008), 'CO_2 embodied in international trade with implications for global climate policy', *Environmental Science and Policy*, **42** (5), 1401–1407.
- ***G. Baiocchi and J.C. Minx (2010), 'Understanding changes in the UK's CO_2 emissions: A global perspective', *Environmental Science and Technology*, **44**, 1177–1184.
- ****M.D. Webster et al. (2003), 'Uncertainty analysis of climate change and policy responses', *Climatic Change*, **61**, 295–320.
- ****M.D. Webster and C.-H. Cho (2006), 'Analysis of variability and correlation in long-term economic growth rates', *Energy Economics*, **28** (5–6), 653–666.

NOTES

1 $\dfrac{\partial lnX}{\partial t} = \dfrac{1}{X}\dfrac{\partial X}{\partial t} = \dfrac{\dot{X}}{X}$

2 Of course, 1.6+1.7−1.0−0.1=2.2 rather than 1.9; the difference is there because multiplicative decomposition is imperfect.

3

Abatement costs

TWEET BOOK

- Emission reduction costs money as it forces companies and households to use more expensive energy and dearer technology. #climateeconomics
- The economy can be fully decarbonized at negligible cost if policy design is smart and abatement is gradual. #climateeconomics
- Differences between abatement cost estimates are large, reflecting different assumptions on emissions without policy. #climateeconomics
- Models also differ on the degree of adjustability of the economy, and the responsiveness of R&D to climate policy. #climateeconomics
- The two degrees target may be physically impossible. It is infeasible without stringent policy in all large countries now. #climateeconomics
- An initially low but rising carbon tax stabilizes the climate. A stringent target requires a high initial carbon tax. #climateeconomics
- Capital stock turnover, technological progress, discount rate and carbon cycle argue for a slow start to abatement. #climateeconomics
- Abatement costs money in a perfect market. In an imperfect market, abatement costs money too. #climateeconomics
- Climate policy may reduce market imperfections and this would at least partly offset the cost of abatement. #climateeconomics
- The revenue of a carbon tax or permit auction could be used to improve the structure of the tax system. #climateeconomics
- Taxes are more distortionary if the tax base is narrower, price elasticities higher, and initial tax level higher. #climateeconomics
- In Europe, the tax burden should be shifted from labour to emissions. This may even stimulate growth. #climateeconomics
- Smart climate policy requires a political system capable of delivering smart fiscal reform. This is in short supply. #climateeconomics

3.1 The costs of emission reduction**

Emission reduction costs money. There are various ways to look at this. Without climate policy, greenhouse gas emissions are free. With climate policy, emissions are not. What used to be free is no longer. Therefore, costs have gone up. Alternatively, you can look at this mathematically. Climate policy imposes a new constraint on a maximization problem. If the constraint bites – that is, if emissions are lower than they otherwise would have been – the objective function must fall. Put yet another way, climate policy forces people and companies to use different technologies and different fuels than they would have without climate policy. Without climate policy, these technologies and fuels are available, but people choose not to use them, or not to the same extent. More specifically, climate policy gets people and companies to invest more in energy savings than they would of their own volition, and gets them to switch to more expensive energy sources. That costs money.

As with any other policy, it is difficult to estimate the costs of climate policy. Most climate policy analysis is done *ex ante* – before the fact. We study a hypothetical situation, or rather, two hypothetical situations, as a cost estimate is the difference in welfare with and without the policy. If we evaluate the impact of past policy, we observe only one history. The "history" without policy is a counterfactual – what would have happened if. Cost estimates therefore must rely on models. We compare two model runs for *ex ante* policy analysis, and we compare a model run with reality for *ex post* policy evaluation.[1] Cost estimates are only as good as the models used.

Not all models are equally good. Some analysts claim that all investments in energy savings are because of climate policy – although in fact energy efficiency has always been improving, well before the advent of climate policy (see Figure 2.2). Famously, George W. Bush once promised to improve energy efficiency by 14% over a decade – even though the historical trend is 18% per decade in the USA. Similarly, other people claim that all investment in renewable energy can be ascribed to climate policy. Truth is, renewable energy is commercially viable in a number of niche applications. Solar power, for instance, beats other sources of electricity if the distance to the grid is sufficiently large.

Estimates of emission reduction costs vary widely, partly because all estimates are model based, and partly because there is little existing climate policy to calibrate models to. Most studies agree, however, that a complete decarbonization of the economy can be achieved at a reasonable cost if policies are

Table 3.1 The total costs of greenhouse gas emission reduction

Target	650 ppm		550 ppm				450 ppm			
Approach	below		above		below		above		below	
Non OECD	now	later	now	later	now	later	now	later	now	later
Model 1	−0.2	0.5	4.8	6.4	5.1	7.4	36.2	78.6	54.4	X
Model 2	13.4	18.8	30.4	48.2	30.9	64.1	123.4	X	X	X
Model 3	23.8	18.9	33.9	26.3	38.0	X	56.7	X	X	X
Model 4	1.4	1.2	3.8	5.1	5.1	10.2	X	X	X	X
Model 5	15.6	17.3	29.7	X	32.7	X	X	X	X	X
Model 6	7.2	7.8	16.2	29.8	18.8	35.7	X	X	X	X
Model 7	2.2	6.5	4.4	9.1	10.9	X	11.9	X	X	X
Model 8	2.2	na	5.9	na	12.4	na	27.9	X	X	X
Model 9	2.4	3.1	5.3	6.7	6.5	X	15.5	32.8	25.7	X
Model 10	13.0	12.8	44.3	59.8	44.3	59.8	X	X	X	X
Model 11	1.9	2.6	27.9	39.7	32.1	64.5	X	X	X	X

Notes:
Costs are the net present value of the abatement costs over the 21st century. Costs are given in trillions of dollar. Results are presented for 11 different models and model variants, and for 3×2×2 policy scenarios with different stabilization targets (in parts per million of carbon dioxide equivalent), different approaches to those targets (from below, that is the target caps concentrations at all times, or from above, that is the target holds for 2100 but may be exceeded in the interim), and for different participation (non-OECD countries start to reduce their emissions in the near future or later in the century). Infeasible scenarios are marked X.

Source: L.E. Clarke, J.A. Edmonds, V. Krey, R.G. Richels, S.K. Rose and M. Tavoni (2009), 'International Climate Policy Architectures: Overview of the EMF22 International Scenarios', *Energy Economics*, **31** (S2), S64–S81.

smart, comprehensive and gradual. These conditions are further discussed below and in Chapter 4.

Models disagree, however, on how much emission reduction would cost. This is illustrated in Table 3.1: Emission reduction costs vary by an order of magnitude. There are various reasons for this. Modellers make different assumptions about what options are available to reduce greenhouse gas emissions, and at what cost. Obviously, if a model omits an option – say, hydrogen fuel-cells for private transport – or assumes that its costs are high,[2] then that model will find that emission reduction is more expensive. Vice versa, if a model assumes that an option exists – say, unlimited capacity for carbon storage – or puts its costs at a lower level than what is commonly believed, then that model will find that emission reduction is less expensive.

The rate of technological change is a key determinant of future emission reduction costs. The difference in the costs between carbon-neutral energy (solar, wind, nuclear) and carbon-emitting energy (coal, oil, gas), for instance, is a key assumption: Emission reduction would be cheap if solar is only slightly more expensive than coal. That cost difference is reasonably well known for the present and past, but has to be assumed for the future. If technology advances faster in carbon-neutral energy than in carbon-emitting energy – say, solar is getting cheaper faster than coal – abatement costs are lower. Different models make different assumptions about the rates of technological progress.

Some models assume that progress in carbon-saving technologies accelerates in response to climate policy. Other models do not have such a response. Of the models that do, some assume that there is no opportunity cost to accelerating technological progress in energy; others do include an opportunity cost. These assumptions further explain the wide range in cost estimates.

If a model assumes high price elasticities, high substitution elasticities, and rapid depreciation of capital, its cost estimates will be lower than those of a model with low price elasticities, low substitution elasticities, and slow turn-over of the capital stock. The latter model assumes that the world of energy use is set in its carbon-intensive ways, which makes it hard and expensive to change course.

Finally, some models assume that, in the scenario without climate policy, greenhouse gas emissions will not grow very fast. Consequently, emission targets are within easy reach. Other models assume rapidly rising emissions, so that a large effort is needed to meet emissions targets.

Table 3.1 shows results for different policy scenarios. There is one minor variation: Is the long-term target an upper bound for the concentrations in all years, or only in the final year? This makes a difference in any model, as the latter case has fewer constraints than the former case. However, there is so much momentum in both the carbon cycle and the energy system that the difference is small. Besides, you would have to rely on the natural processes in the carbon cycle to remove the excess carbon dioxide from the atmosphere. That puts a limit on the extent of the overshoot: In most cases, it is optimal to approach the target from below.

Some of the models in Table 3.1, however, assume that biomass power plus carbon capture and storage is a viable option at scale. This is negative-carbon-energy: Plants take up carbon dioxide when growing. In a normal biomass power plant, roughly the same amount of carbon dioxide is released again. But if the carbon is captured and stored, the net effect is that you remove carbon dioxide from the atmosphere. This implies that you can correct excess emissions from earlier years towards the end of the century. In this case, you can overshoot the target in the intermediate years to a larger extent, and costs are saved.

Participation of poorer countries in climate policy is another variation in the policy scenarios shown in Table 3.1. In some scenarios, every country starts to reduce its emissions from 2015 onwards. In other scenarios, only rich countries do, and poorer countries start considerably later. This has a large impact on the estimated cost of emission reduction. If a fraction of emission is excluded from abatement, the rest will have to be reduced more to meet the same target. As emission reduction costs are more than linear in emission reduction effort, this necessarily drives up the total costs. Furthermore, many of the cheaper emission reduction options can be found in poorer countries, partly because these economies tend to rely on older, less efficient technology, and partly because money buys more in poorer countries.

The concentration target is the third policy variation in Table 3.1. The more stringent the target, the higher the cost – and costs rise very rapidly from the more lenient to the more ambitious targets. For the most stringent targets, a number of models do not report. That can be for one of three reasons. First, the representation of the carbon cycle disallows the model to meet the target. Second, the representation of emissions and emission reduction disallows the target. Third, the model can meet the target, but the costs are so exorbitant that the modeller refused to report the results. Whatever the reason – physical, technical or political – the most stringent target in Table 3.1 may well be beyond reach. This is as expected: There are always things than cannot be done. However, the 450 ppm CO_2eq target in Table 3.1 corresponds to a 50–50 chance of meeting the 2°C target of the European Union and the United Nations.

Table 3.2 complements Table 3.1. It shows results for the same set of models and the same set of scenarios, but now for the marginal abatement costs. This is best thought of as the carbon tax imposed on all greenhouse gas emissions from all economic activities in all (participating) countries in 2015. Per policy scenario, the models again disagree by an order of magnitude. The initial carbon tax required for meeting the least stringent target is modest, but this escalates with increased stringency.

Table 3.2 shows the increase in energy prices in dollars per tonne of carbon dioxide, a unit with which not everyone is intimately familiar. Therefore, Table 3.3 translates $/t$CO_2$ into local currency per unit of energy use. That is, Table 3.3 specifies how much a carbon tax would add to a liter of gasoline, a bag of coal, and a kilowatt-hour of electricity.

Table 3.2 The marginal costs ($/tCO$_2$eq) of greenhouse gas emission reduction

Target	650 ppm				550 ppm				450 ppm	
Approach	below		above		below		above		below	
Non OECD	now	later	now	later	now	later	now	later	now	later
Model 1	3	5	8	13	10	24	77	214	1297	X
Model 2	20	43	51	147	52	239	260	X	X	X
Model 3	14	16	27	28	27	X	28	X	X	X
Model 4	1	1	11	12	16	92	X	X	X	X
Model 5	13	27	43	X	52	X	X	X	X	X
Model 6	9	13	29	154	35	256	X	X	X	X
Model 7	6	35	7	35	26	X	15	X	X	X
Model 8	6	na	12	na	27	na	70	X	X	X
Model 9	4	7	8	10	14	X	20	53	101	X
Model 10	10	11	40	67	30	67	X	X	X	X
Model 11	3	6	4	36	22	131	X	X	X	X

Notes:
Marginal costs are for 2020, and apply (uniformly) to the participating countries only. Marginal costs are given in dollars per tonne of carbon dioxide equivalent. Results are presented for 11 different models and model variants, and for 3×2×2 policy scenarios with different stabilization targets (in parts per million of carbon dioxide equivalent), different approaches to those targets (from below, that is the target caps concentrations at all times, or from above, that is the target holds for 2100 but may be exceeded in the interim), and for different participation (non-OECD countries start to reduce their emissions in the near future or later in the century). Infeasible scenarios are marked X.

Source: L.E. Clarke, J.A. Edmonds, V. Krey, R.G. Richels, S.K. Rose and M. Tavoni (2009), 'International Climate Policy Architectures: Overview of the EMF22 International Scenarios', *Energy Economics*, **31** (S2), S64–S81.

Table 3.2 shows the required carbon tax in 2020. The carbon tax is assumed to increase over time. Figure 3.3 provides insight into the allocation of emission reduction effort over time. Figure 3.3 shows the emissions trajectories to meet five alternative targets at the lowest possible costs. It contrasts the least cost trajectories to arbitrary trajectories for four of the five targets. Figure 3.4 shows the cost differences, which vary between 10% and 60% depending on the model. As the net present value of the costs of emission reduction is measured in trillions of dollars, a 10% cost savings is worth pursuing. The main difference between the two sets of trajectories is that the arbitrary ones start with radical emission cuts whereas the least cost trajectories begin with modest abatement that accelerates over time.

There are four reasons why money is saved if emission reduction targets are lenient at first. Emission reduction requires changes in behaviour and technology. Behaviour and technology, however, are constrained by durable consumption goods and invested capital. A carbon tax does not reduce the

Table 3.3 Carbon dioxide emissions per unit of energy use and price increase due to a $100/tC carbon tax

Fuel	Unit	Brazil	China	Germany	France	India	Japan	UK	USA
				Emissions per unit					
Petrol	kgCO$_2$/l	2.312	2.312	2.312	2.312	2.312	2.312	2.312	2.312
Diesel	kgCO$_2$/l	2.668	2.668	2.668	2.668	2.668	2.668	2.668	2.668
Gas	kgCO$_2$/kWh	0.184	0.184	0.184	0.184	0.184	0.184	0.184	0.184
Coal	kgCO$_2$/kg	2.383	2.383	2.383	2.383	2.383	2.383	2.383	2.383
Power	kgCO$_2$/kWh	0.076	0.794	0.451	0.097	1.239	0.437	0.487	0.544
				Carbon tax[a]					
Carbon tax	LC/tCO$_2$	64	168	21	21	1784	2715	17	27
Carbon tax	LC/tC	235	617	76	76	6540	9955	64	100
				Price increase per unit[a]					
Petrol	LC/l	0.148	0.389	0.048	0.048	4.123	6.276	0.040	0.063
Diesel	LC/l	0.171	0.449	0.055	0.055	4.758	7.243	0.047	0.073
Gas	LC/kWh	0.012	0.031	0.004	0.004	0.327	0.498	0.003	0.005
Coal	LC/kg	0.153	0.401	0.049	0.049	4.250	6.470	0.042	0.065
Power	LC/kWh	0.004	0.125	0.009	0.002	1.697	1.126	0.008	0.014

Note: [a] LC = local currency: real, renminbi, euro, euro, rupiah, yen, pound sterling, dollar.

emissions of those households and companies that continue to use the same cars, live and work in the same place and in the same building, and operate the same machinery. In those cases, a carbon tax simply imposes a penalty on investment decisions made in earlier, pre-climate-policy times. This is a deadweight loss to the economy. This deadweight loss falls over time as capital turns over, so that the carbon tax can increase with inducing excessive costs.

Technological change is another reason why emission reduction is expensive in the short term but cheaper in the medium to long term. Carbon-neutral energy is still immature technology. Although fossil fuel technology continues to progress, it is well developed and all the easy improvements have been made. Although there has been rapid progress in oil and gas exploitation, this has been about unlocking relatively expensive reserves. In contrast, we can still expect major technological breakthroughs with solar power and bioenergy. Furthermore, the easily accessible sources of fossil fuels are getting exhausted. So, over time, we expect the costs of fossil fuels to rise and the costs of renewables to fall. As the costs of emission reduction are driven by the difference in costs between fossil and renewable energy, abatement costs should fall over time.

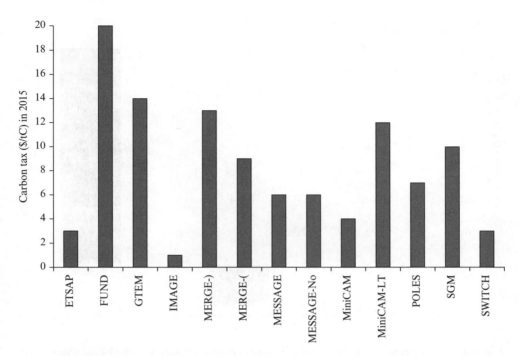

Figure 3.1 The marginal costs of emission reduction in 2015 needed to meet, with full participation, a 650 ppm CO_2eq target in 2100 according to different models

Third, emission reduction costs in the future are discounted. Postponing emission reduction reduces the net present value of the costs. Fourth, emissions are degraded in the atmosphere. Climate policy targets typically refer to the long term, say the year 2100. Emissions in 2090 are more important to concentrations in 2100 than emissions in 2015. Atmospheric degradation thus functions as a discount rate, so that it is better to reduce emissions later.

3.2 Negative abatement costs**

There are claims that the costs of emission reduction are negative – that is, that it would be possible to save emissions and save money at the same time. Some of the claims are the result of bad accounting. Two common mistakes are the following. First, people confuse the technological change that is part of the no-policy scenario with the accelerated technological change in the policy scenario. As we saw in Chapter 2, the no-policy scenario indeed contains a large number of actions that are both commercially viable and reduce emissions. Energy efficiency improves over time, also in the absence of climate policy. Because these investments are commercially viable, they do

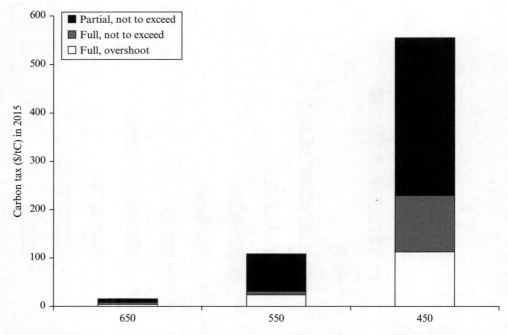

Figure 3.2 The marginal costs of emission reduction in 2015, averaged across models, needed to meet alternative targets in 2100 (or throughout the 21st century) with different participation rates

not need policy support – and it is thus wrong to attribute them to climate policy.

Another common mistake is to underestimate the costs of investment. For example, some analysts assume that households and companies can borrow money at the same rate of interest as the government can. In fact, private rates of interest tend to be higher than public ones. That makes investment less attractive. As another example, well-established technologies have acquired a reputation and a dense network of mechanics for installation, maintenance and repair. New technologies lack those, a cost that is easily overlooked.

Non-economists may also claim that a reduction in fossil fuel imports would be good for the economy. Jean-Baptise Colbert was an early proponent of import substitution as a strategy for economic growth – mercantilism – but the theory was discredited a long time ago. Import substitution policies were largely abandoned in the 1980s. Substituting cheap imported energy with expensive domestic energy slows economic growth. Protected infant industries tend not to create competitive companies, but rather companies that are adept at lobbying and rent-seeking. The balance of payments holds, of

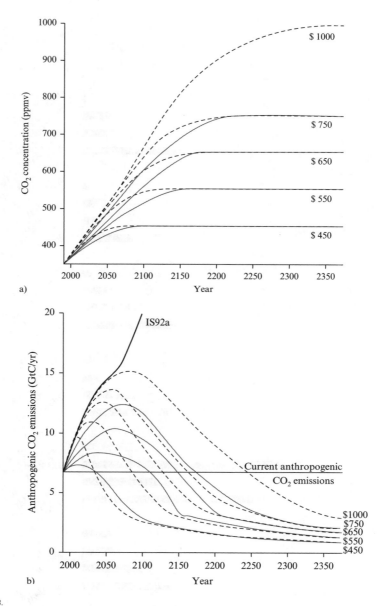

Source: IPCC WG3 AR3.

Figure 3.3 Alternative pathways to stabilization of carbon dioxide concentrations in the atmosphere

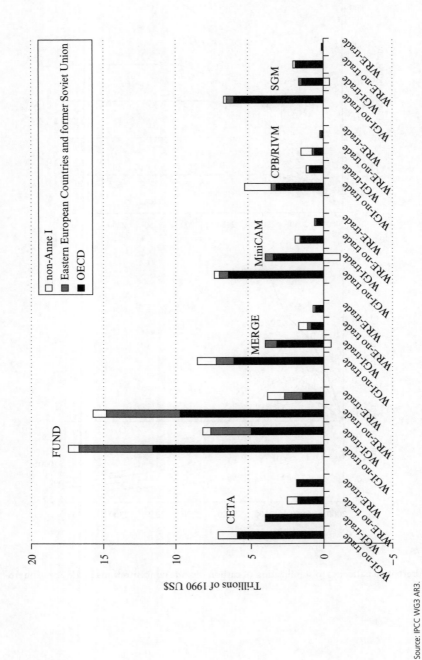

Source: IPCC WG3 AR3.

Figure 3.4 The costs of alternative pathways to stabilization of carbon dioxide concentrations in the atmosphere; WRE corresponds to the black lines in Figure 3.3, WGI to the blue lines

Table 3.4 The costs of emission reduction in USA according to four models, for alternative carbon tax revenue recycling options

	Model 1	Model 2	Model 3	Model 4
Lump sum transfer to households	−0.58	−0.46	−0.62	−0.24
Increase government spending	−0.40	−1.02		−0.24
Reduce personal income tax	−0.56	−0.53	−0.16	−0.16
Reduce corporate income tax	0.40	−0.11	0.60	−0.17
Reduce payroll tax				−0.18
Reduce payroll tax paid by employer	−0.58	−0.53		
Reduce payroll tax paid by employee	0.19	−0.25		
Increase investment credit	1.55	1.67		0.00

course, so reduced imports imply reduced exports, reduced foreign investment, exchange rate adjustments and so on.

That said, there may be genuine reasons why the costs of emission abatement may be different than suggested by Tables 3.1 and 3.2 – perhaps smaller or even negative. The models in these tables are either optimization models or equilibrium models. Recall that a market equilibrium corresponds to a Pareto optimum. If the no policy scenario is an optimum, any policy intervention bears a cost. If you start at the top, the only way is down.

In reality, however, the no-climate-policy case is characterized by many market imperfections and policy distortions. Climate policy may overcome some of these, and this would reduce its costs. However, climate policy may also interact with pre-existing distortions, and this would increase its costs.

A carbon tax is one way to implement climate policy. Like any tax, a carbon tax is distortionary. In an undistorted market, rational actors find a Pareto optimum. A tax changes the choices people make, and leads that market to an equilibrium with lower welfare. The welfare loss is a measure for the degree of distortion of the tax.

However, a carbon tax brings revenue too, and that revenue could be used to reduce other, more distortionary taxes. Taxes are distortionary because they distort behaviour, moving people and companies away from the optimum. Taxes are more distortionary if they are higher, if price elasticities are higher (because behaviour is more responsive), and if the tax base is narrower (as fewer people are affected, by definition, then, for the same revenue, the behaviour of those people is further distorted). A carbon tax starts from a low level, price elasticities are low, and a carbon tax has a broad base. It is therefore not particularly distortionary (even though it is specifically designed to change behaviour). If the carbon tax revenue is used to reduce another tax, there may well be a benefit – and that benefit may more than offset the initial cost of abatement.

Online Figure 3.1 illustrates this, comparing three alternative welfare measures and twelve European countries for a single carbon tax and a single carbon-tax recycling scheme: the reduction of payroll taxes. In the majority of cases, welfare increases. If payroll taxes fall, companies would hire more workers. Online Figure 3.2, which is taken from the same study, confirms this. However, these benefits are not automatic. Online Figure 3.3 shows the results for the same carbon tax again but different recycling options. Depending on the country (or rather, its pre-existing fiscal policy), revenue recycling brings larger or smaller benefits. Comparing Online Figure 3.1 and Online Figure 3.3, you may conclude that a payroll tax reduction is best. Table 3.4 shows that that conclusion is unfounded. Table 3.4 shows results for the USA. In Europe, labour taxes tend to be high and are thus a prime target for a beneficial reduction. In the USA, tax reform that stimulates savings and investment is more desirable.

In sum, the revenue of a carbon tax may be used to reduce other taxes and this would bring benefits that at least partially offset the costs of emission reduction. If the tax reform is well-tailored to the specific circumstances of the fiscal system, then that benefit may be substantial. It is not the case that any use of the revenue is beneficial: It may be used to increase hand-outs to friends and allies of the government. It is also not the case that any tax reform is equally beneficial. The benefits that exist in theory are not necessarily realized in practice.

 FURTHER READING

Every six years, Working Group III of the Intergovernmental Panel on Climate Change publishes a major assessment of the costs of emission reduction. The information is layered, with a Summary for Policy Makers with high-level information, Technical Summaries with more detail, and multiple chapters with a lot of detail and references to the underlying literature. These reports can be found at: http://www.ipcc.

The Energy Modelling Forum regularly organizes model comparison exercises on abatement costs. Recent and relevant are EMF25, EMF22 and EMF21, which can be found at http://emf. stanford.edu/research/.

IDEAS/RePEc has a curated bibliography on this topic at http://biblio.repec.org/entry/tdb.html.

? EXERCISES

3.1 Table 3.1 shows the total cost of greenhouse gas emission reduction. Calculate the average across the 11 models. Calculate the average extra costs for delayed participation by non-OECD countries. Calculate the average extra costs for approaching the target from below. Calculate the average extra costs of making the target more stringent by 100 ppm.

3.2 Assume that emission reduction costs are exponential in relative emission reduction $C = \alpha \exp(R)$, where C denote costs, R relative emission reduction, and $\alpha = 1$ is a parameter. Suppose that the emissions target is $T = (1-R) E$, where $E = 100$ are baseline emissions. Compute the costs of emission reduction. Compute the change in costs of emission reduction if the cost parameter is 10% higher, i.e., $\alpha = 1.1$. Compute the change in costs if baseline emissions are 10% higher, i.e., $E = 110$.

3.3 The results in Table 3.1 vary widely. How would you go about testing which model is correct?

3.4 Read and discuss:

- **M. Wise, K. Calvin, A. Thomson, L. Clarke, B. Bond-Lamberty, R. Sands, S.J. Smith, A. Janetos and J. Edmonds (2009), 'Implications of limiting CO_2 concentrations for land use and energy', *Science*, **324**, 1183–1186.

- **T.C. Schelling (1996), 'The economic diplomacy of geoengineering', *Climatic Change*, **33**, 303–307.

- ***S. Barrett (2008), 'The incredible economics of geoengineering', *Environmental and Resource Economics*, **39**, 45–54.

- ***G.C. van Kooten, A.J. Eagle, J. Manley and T. Smolak (2004), 'How costly are carbon offsets? A meta-analysis of forest carbon sinks' *Environmental Science and Policy*, **7**, 239–251.

- ****M. Goes, N. Tuana and K. Keller (2011), 'The economics (or lack thereof) of aerosol geoengineering', *Climatic Change*, **109**, 719–744.

- ****R.N. Lubowski, A.J. Plantinga and R.N. Stavins (2006), 'Land-use change and carbon sinks: Econometric estimation of the carbon sequestration supply function', *Journal of Environmental Economics and Management*, **51**, 135–152.

- ****L.H. Goulder (1995), 'Environmental taxation and the double dividend: A reader's guide', *International Tax and Public Finance*, **2**, 157–183.

- ****S.A. Smulders and M. de Nooij (2003), 'The impact of energy conservation on technology and economic growth', *Resource and Energy Economics*, **25**, 59–79.

NOTES

1 I am not aware of any *ex post* estimate of the costs of greenhouse gas emission reduction. There are many studies of the European Emissions Trading System for carbon dioxide emission permits, but these focus on the market rather than the welfare implications. There are a number of studies on particular aspects of climate policy, such as to what extent it accelerates or redirects technological progress, or on particular policy interventions in particular countries. There is no comprehensive estimate, however.

2 Of course, assuming a prohibitively high cost is equivalent to assuming unavailability.

4

Policy instruments for emission reduction

- Direct regulation is the government telling people and companies what (not) to do, and how (not) to do it. #climateeconomics
- Direct regulation works fine if there are few, similar sources of emissions. It does not work for greenhouse gases. #climateeconomics
- Direct regulation was successful, it made the career of current environmental leaders, and so it is still popular. #climateeconomics
- Taxes, subsidies and tradable permits reward emission reduction, but the decision whether and how is private. #climateeconomics
- Taxes increase costs, subsidies reduce costs. Over time, subsidies thus lead to an expansion of the polluting sector. #climateeconomics
- With permit trade, emissions are known but costs are not. With taxes, marginal costs are known but emissions are not. #climateeconomics
- (Mistakes with) annual emissions do not matter much for a stock pollutant. Emission certainty is worth little. #climateeconomics
- Cost certainty is worth a lot. Therefore, taxes are better suited for climate policy than tradable permits. #climateeconomics
- There are various ways to allocate permits: auction, equal per capita, to the polluters, to the victims of pollution. #climateeconomics
- The initial allocation of permits provides opportunities for politicians to hand out favours. #climateeconomics
- The willingness to pay to pollute equals the willingness to accept compensation for not polluting. #climateeconomics
- The willingness to accept compensation for pollution equals the willingness to pay for reduced pollution. #climateeconomics
- The market allocation of permits is therefore independent of the initial allocation. Efficiency and equity separate. #climateeconomics
- The EU ETS is the largest and only international emission permit market. It suffered from avoidable teething problems. #climateeconomics
- Enforcement is the greatest challenge for the EU ETS, as it relies on judicial strength of individual Member States. #climateeconomics

> - Technological progress in renewable energy, energy efficiency and agriculture drives the costs of climate policy. #climateeconomics
> - Diverting R&D towards energy and agriculture would be expensive as these are small sectors in the economy. #climateeconomics
> - R&D is best stimulated by patents, prizes, and taxes. Governments are bad at picking winners. #climateeconomics

4.1   The justification of public policy*

The First Welfare Theorem has that a competitive equilibrium is a Pareto optimum. The intuition is as follows. A voluntary exchange is Pareto improving: Both parties are at least as well off as without the exchange. Why else would they agree to it? A sequence of voluntary exchanges thus improves welfare. If there is no additional exchange possible that satisfies all parties, then we must be in a Pareto optimum – but the market must also be in equilibrium as no further exchanges take place.

The First Welfare Theorem can be used to argue that the government should leave the market well alone, as any intervention would be Pareto inferior. There are a number of exceptions to this. If there are externalities, the market equilibrium is not a Pareto optimum. The intuition is simple. An externality is an unintended and uncompensated impact. If two parties voluntarily agree on an exchange, that exchange must be Pareto improving. However, if this exchange unintentionally hurts a third party, and the two exchange parties do not make good the damage (nor cancel the exchange), then the exchange is no longer Pareto superior. A sequence of such exchanges would not lead to a Pareto optimum.

Carbon dioxide clearly is an externality. We burn fossil fuel to generate electricity and propel cars. We do not burn fossil fuel to emit carbon dioxide. Emissions are thus unintentional (even if intrinsic to the process; see Section 2.2). Climate change does affect the welfare of people all over the world. These people are not compensated by the emitters of carbon dioxide.

In the presence of externalities, government intervention can improve welfare and is thus justified. The best intervention is the Pigou tax, named after Arthur Pigou. The Pigou tax does three things. First, it puts a tax on the activity that generates the externality. Second, it uses the tax revenue to compensate the victims of the externality.[1] Third, the compensation is such that it offsets the loss of welfare at the margin.

4.2   Direct regulation*

The regulator has many ways to affect emissions. Each of these instruments has different properties, which makes them more suitable for solving some problems than others.

Direct regulation is probably the most common form of environmental policy. Direct regulation has been highly successful in the OECD. In the 1960s and 1970s, the environment in Europe and North America was filthy. It is no longer. The clean-up of the environment was done by direct regulation. This means that environmental regulators have a substantial amount of experience with these instruments, while senior regulators look back at the past successes of direct regulation. Even though times have changed and current environmental problems are different, environmental regulation lags behind. Furthermore, direct regulation allows bureaucrats to expand bureaucracies.

Direct regulation is also known as command and control. Essentially, the regulator goes in and tells household and companies what (not) to do and how (not) to do it. The regulator would be able to come up with sensible instructions if she has detailed knowledge of the regulated activity, which requires that there are either only a small number of agents or a small number of technologies in use. Direct regulation is essentially a one-size-fits-all solution. Regulation is homogenous because of capacity constraints within the regulator, and because administrative fairness demands that everyone is treated the same. This is fine unless there is substantial heterogeneity among the regulated.

There are different forms of direct regulation:

- The regulator may proscribe or forbid certain inputs into the production process, or put standards on the amount of input used.
- The regulator may proscribe or forbid certain technologies used in the production process, or put standards on performance.
- The regulator may put limits on selected outputs of the production process, or put requirements on the products.
- The regulator may put limits on the timing of certain activities, or on their location.

For example, government have mandated that car fuel should be a blend of petrol and bioethanol. Car fuel may not contain lead. Car engines have to be equipped with catalytic convertors and meet fuel efficiency standards.

Power plants may only emit a certain amount of sulphur. Toys for infants may not contain carcinogenic material. Planes are not allowed to take off or land between 11 pm and 6 am. New buildings cannot be built in nature reserves. CFCs may not be made, sold, bought or used.

4.3　Market-based instruments*

Market-based or incentive-compatible policy instruments are the main alternative to direct regulation. Taxes and subsidies are the oldest instruments. With a tax, there is a charge, levy or penalty for every unit of the offending substance (or a proxy) used, produced, or emitted. With a subsidy, there is monetary reward for every unit of the offending substance not used, not produced, or not emitted.

In the short run, taxes and subsidies have the same effect on, say, emissions. With a subsidy, every tonne of emissions avoided will bring a reward. With a tax, every tonne of emissions avoided will reduce the tax burden, that is, bring a reward.

Taxes and subsidies have different distributional effects. With a tax, money flows from households and companies to the government. With a subsidy, money flows from the government to households and companies.

Because of that, taxes and subsidies also have different effects on emissions in the medium run. An emission tax increases the average cost of doing business in a particular sector. Investment flows elsewhere and the emitting sector shrinks (relative to what its size would have been without the tax). An emission avoidance subsidy reduces the average cost of doing business in that sector. Additional investment flows there, and the emitting sector expands (relative to what its size would have been without the subsidy).

Tradable permits are a more recent addition to the set of instruments available to the regulator. With tradable permits, the regulator sets an overall cap on consumption, production, or emissions. Let us focus on the last. The overall emissions cap is then split into units and each emitter receives a certain amount of permits to emit. So far, this is direct regulation. However, if a company finds that it has too few permits, it may buy additional permits from a company that has too many.

The price for emission permits that is formed on its market works just like a tax. For every unit of additional emissions, a company either has to buy an additional permit (which is a cost) or can sell fewer of the permits it holds (which is a cost too). For every unit of emissions avoided, a company either can sell more permits (which is a benefit) or has to buy fewer permits in the market (which is a benefit too).

The main advantage of market-based instruments is that the regulator does not specify *how* emissions are reduced. That decision is left to household and companies. The regulator does specify, however, *that* emissions are reduced.

4.4   Cost-effectiveness*

Furthermore, the costs of emission reduction are uniform at the margin. This is an important characteristic. Let us consider a social planner, who seeks to reduce emissions at a minimum cost to society:

$$C = \sum_n C_n = \sum_n \alpha_n M_n + \beta_n M_n^2 \qquad (4.1)$$

where C are the social costs, C_n are the costs per company n, M_n are the emission reduction efforts of company n, and α and β are parameters. Let M denote the desired total emission reduction effort. Then, the least-cost emission solution follows from

$$\min_{M_n} \sum_n C_n \text{ s.t. } \sum_n M_n > M \qquad (4.2)$$

Form the Lagrangian

$$L = \sum_n \alpha_n M_n + \beta_n M_n^2 - \lambda \left(\sum_n M_n - M \right) \qquad (4.3)$$

and take the first partial derivative to the policy instruments (i.e., the emission reduction effort) to derive the first-order conditions for optimality:

$$\frac{\partial L}{\partial M_n} = \alpha_n + 2\beta_n M_n - \lambda = 0 \; \forall n \Rightarrow \frac{\partial C_n}{\partial M_n} = \lambda \forall n \qquad (4.4)$$

That is, least-cost emission reduction requires that all emitters face the *same* abatement cost at the margin. Because there is a shared constraint M, the shadow price of the constraint λ is set at the societal level and is thus the same for all emitters.

The least-cost solution to meet a target is known as the cost-effective solution. Cost-efficacy is an optimum. A solution cannot be more cost-effective than another solution; it either is cost-effective or it is not. Some people use the words "more cost-effective" as an "erudite" alternative to the word "cheaper", but in fact they demonstrate their lack of understanding of the meaning of the concept cost-efficacy. Other people, particularly native speakers of French and German, use the word "cost-efficiency" as a synonym for cost-efficacy. In fact, cost-efficiency is the dual of production-efficiency, and if you do not understand what that means, then you should not use the word cost-efficiency.

Now let us consider a company faced with an emissions tax. It seeks to minimize its costs

$$\min_{M_n} \alpha_n M_n + \beta_n M_n^2 - t M_n \forall n \tag{4.5}$$

The cost function is as in Equation (4.1), but for every unit of emission reduction effort M, it pays t less in tax.

Equation (4.5) is an unconstrained optimization problem, so the first-order condition has that the first partial derivative equals zero:

$$\alpha_n + 2\beta_n M_n - t = 0 \quad \forall n \Rightarrow \frac{\partial C_n}{\partial M_n} = t \quad \forall n \tag{4.6}$$

Equation (4.6) is identical to Equation (4.4) if $t = \lambda$.

If the regulator uses tradable permits, Equation (4.5) becomes

$$\min_{M_n} \alpha_n M_n + \beta_n M_n^2 - p M_n \forall n \tag{4.7}$$

where p is the permit price.

If the regulator uses subsidies, Equation (4.5) becomes

$$\min_{M_n} \alpha_n M_n + \beta_n M_n^2 + s M_n \quad \forall n \tag{4.8}$$

where s is the subsidy.

That is, a uniform emission tax, a uniform emission avoidance subsidy, and an emission permit market with a uniform price all lead to uniform marginal abatement costs. Put differently, taxes, subsidies, and emission permits guarantee cost-effectiveness.

There is no such guarantee for direct regulation. In fact, the regulator would need to know the marginal abatement cost function of each of the regulated households and companies in order to achieve cost-efficacy. That is unrealistic unless there are few agents or all agents use the same technology in the same way.

Uniformity can be deceptive. There are two types of combustion engine: Diesel and Otto. Suppose that the government imposes minimum fuel efficiency on petrol cars. This makes cars more expensive. A long-distance commuter would avoid a lot of carbon dioxide emissions. Someone who drives to church once a week would avoid few emissions. There is thus a large difference in average costs, even if the regulation is seemingly uniform.

4.5 Dynamic efficiency****

Above, we derive the condition for static efficiency: A uniform carbon price. Below, we derive the conditions for dynamic efficiency under three alternative problem definitions.

4.5.1 Emission reduction as a resource problem

Climate policy can be looked at as a waste disposal problem. There is some finite disposal capacity, and every emission degrades some of that capacity. Cutting emissions, though, affects output. The problem can then be formalized as follows. Let us maximize net present welfare:

$$\max_{C(t),E(t)} W = \int_t U(C(t))e^{-\rho t}\,dt \text{ for } M > 0 \text{ and } U = \underline{U} \text{ for } M \leq 0 \quad (4.9)$$

subject to

$$\dot{K} = Y(t) - C(t) = Y(K(t), E(t)) - C(t) \quad (4.10)$$

and

$$\dot{M} = \delta M(t) - E(t) \quad (4.11)$$

where W is net present welfare, U is instantaneous utility, C is consumption, Y is output, K is capital, E is emissions and M is total allowable emissions.

The current-value Hamiltonian is

$$H = U(C(t)) + \kappa\dot{K} + \mu\dot{M} = U(C(t)) + \kappa(Y(K(t), E(t)) - C(t)) +$$

$$\mu(\delta M(t) - E(t)) \quad (4.12)$$

The first-order conditions are

$$\frac{\partial H}{\partial C} = \frac{\partial U}{\partial C} - \kappa = 0 \Rightarrow U_C = \kappa \tag{4.13}$$

$$\frac{\partial H}{\partial E} = \kappa \frac{\partial Y}{\partial E} - \mu = 0 \Rightarrow Y_E = \frac{\mu}{\kappa} \tag{4.14}$$

$$\dot{\kappa} = \rho\kappa - \frac{\partial H}{\partial K} = \rho\kappa - \kappa \frac{\partial Y}{\partial K} \Rightarrow \frac{\dot{\kappa}}{\kappa} = \rho - Y_K \tag{4.15}$$

$$\dot{\mu} = \rho\mu - \frac{\partial H}{\partial M} = \rho\mu - \mu\delta \Rightarrow \frac{\dot{\mu}}{\mu} = \rho - \delta \tag{4.16}$$

That is, marginal utility equals the shadow price of capital or the return on savings; see Equation (4.13). The marginal cost of emission reduction (in money) equals the shadow price of the emission allowance (in utile, normalized by marginal utility to convert to money); see Equation (4.14). The growth rate of the shadow price of capital is the difference between the pure rate of time preference and the return to capital; see Equation (4.15).

The growth rate of the shadow price of the emission allowance is the discount rate minus the rate at which waste disposal capacity is added; see Equation (4.16). That is, the carbon tax should grow at the discount rate minus the enhancement rate.

4.5.2   Emission reduction as an efficiency problem

Climate policy can also be looked at as an efficiency problem. Emissions add to concentrations. Welfare depends on the concentration of greenhouse gases in the atmosphere. Cutting emissions affects output. The problem can then be formalized as follows. Let us maximize net present welfare:

$$\max_{C(t),E(t)} W = \int_t U(C(t), M(t))\, dt \tag{4.17}$$

subject to

$$\dot{K} = Y(t) - C(t) = Y(K(t), E(t)) - C(t) \tag{4.18}$$

and

$$\dot{M} = E(t) - \delta M(t) \tag{4.19}$$

The current-value Hamiltonian is

$$H = U(C(t)) + \kappa \dot{K} + \mu \dot{M}$$
$$= U(C(t), M(t)) + \kappa(Y(K(t), E(t)) - C(t)) + \mu(E(t) - \delta M(t)) \tag{4.20}$$

The first-order conditions are (4.13), (4.15) and

$$\frac{\partial H}{\partial E} = \kappa \frac{\partial Y}{\partial E} + \mu = 0 \Rightarrow Y_E = -\frac{\mu}{\kappa} \tag{4.21}$$

$$\dot{\mu} = \rho\mu - \frac{\partial H}{\partial M} = \rho\mu - \frac{\partial U}{\partial M} + \mu\delta \Rightarrow \frac{\dot{\mu}}{\mu} = \rho + \delta - \frac{U_M}{\mu} \tag{4.22}$$

That is, marginal utility equals the shadow price of capital or the return on savings (as above). The marginal cost of emission reduction (in money) equals the shadow price of the emission allowance (in utile, normalized by marginal utility to convert to money); note that the interpretation of the stock equation changed, so that the signed flipped; see Equation (4.21). The growth rate of the shadow price of capital is the difference between the pure rate of time preference and the return to capital (as above).

The growth rate of the shadow price of emissions is the discount rate *plus* the rate of atmospheric degradation (rather than minus as above), and minus the marginal damage of climate change over the shadow price; see Equation (4.22). The marginal damage is given in utils per concentration; the shadow price in utils per emission; the final term on the right-hand side of Equation (4.22) is therefore measured in emission per concentration, that is, a unitless rate like ρ and δ. It measures how rapidly the climate change problem gets worse. Taken together, the shadow price of emissions is higher if we care less about the future, if the future is less problematic because emissions are dissipated, and if the welfare impacts grow less fast.

4.5.3 Emission reduction as a cost-effectiveness problem

Climate policy can also be looked at as a cost-effectiveness problem. Emissions add to concentrations. There is an agreed upper limit on the concentration of greenhouse gases in the atmosphere. This can be formalized as zero damages below the threshold, and arbitrarily high damages above.

Cutting emissions affects output. The problem can then be formalized as follows. Let us maximize net present welfare:

$$\max_{C(t),E(t)} W = \int_t U(C(t), M(t))\, dt \text{ with } \frac{\partial U}{\partial M} = 0 \text{ for}$$

$$M \le \overline{M} \text{ and } U = \underline{U} \text{ for } M > \overline{M} \tag{4.23}$$

subject to (4.18) and (4.19).

The current-value Hamiltonian is (4.20). The first-order conditions are (4.13), (4.15), (4.21) and

$$\dot{\mu} = \rho\mu - \frac{\partial H}{\partial M} = \rho\mu + \mu\delta \Rightarrow \frac{\dot{\mu}}{\mu} = \rho + \delta \text{ for } M \le \overline{M} \tag{4.24}$$

That is, marginal utility equals the shadow price of capital or the return on savings (as above). The marginal cost of emission reduction (in money) equals the shadow price of the emission allowance (as above). The growth rate of the shadow price of capital is the difference between the pure rate of time preference and the return to capital (as above).

The growth rate of the shadow price of the emission allowance is the discount rate *plus* the rate of atmospheric degradation (but without the marginal damage of climate change).

4.5.4 Summary

Comparing the above results, the following insights emerge. If we impose a constraint on concentrations, then the carbon price should rise at the rate of discount *plus* the rate of removal from the atmosphere (until the target is met). If we view greenhouse gas emissions as a waste disposal problem, with a fixed capacity, the same result emerges. Atmospheric degradation adds to the discount rate – or, as formulated above, the carbon price should rise at the rate of discount *minus* the rate of addition to the disposal capacity, which is minus the rate of removal from the atmosphere. If we seek to maximize welfare, the carbon price should rise at the rate of discount plus the rate of atmospheric removal but minus the rate at which the climate problem gets worse.

4.6 Environmental effectiveness*

Environmental effectiveness is the other main criterion for environmental policy. A policy intervention is useless unless it reduces emissions by at least (roughly) the desired amount. Tradable emission permits, if monitored and enforced, guarantee environmental effectiveness, as there is a cap on total emissions.

Taxes and subsidies have no such guarantee. Unless the regulator knows the marginal abatement cost curve of each emitter, the regulator cannot accurately predict how companies and households will respond to either a tax or a subsidy. The regulator therefore does not know by how much emissions will be reduced.

Direct regulation covers many different policy interventions, some of which have relatively certain environmental results. An emission cap per installation is an example, although more installations may be built. Sometimes, emission caps only hold for installations above a threshold size, and companies may opt for more, smaller installations in response.

If direct regulation targets inputs or technologies, behavioural change may partly defy the environmental goals. Improved energy efficiency, for instance, lowers the effective price of energy services and may lead to increased energy use rather than reduced emissions – the so-called rebound effect first discussed by W.S. Jevons in 1865. For instance, building insulation may lead to warmer homes rather than lower energy use.

4.7 Taxes versus tradable permits under uncertainty**

We have seen above that tradable permits lead to a certain environmental outcome, but the permit price and hence the costs of abatement are unpredictable. Vice versa, the marginal cost of abatement is known for an emission tax and this puts an upper bound on the total costs of abatement. However, the effect on the environment is uncertain. The Weitzman Theorem provides guidance which is worse.

Figure 4.1 shows a standard benefit–cost analysis. The marginal costs of abatement rise as emissions are cut further. The marginal benefits of abatement fall with emission reduction. The optimal tax is set where the marginal cost curve crosses the marginal benefit curve. The optimal emission target is set at the same point.

Now suppose that the government believes that emission reduction is more expensive than it really is. If the government uses a quantity instrument, it underregulates: The emissions target is higher than it should be, because the

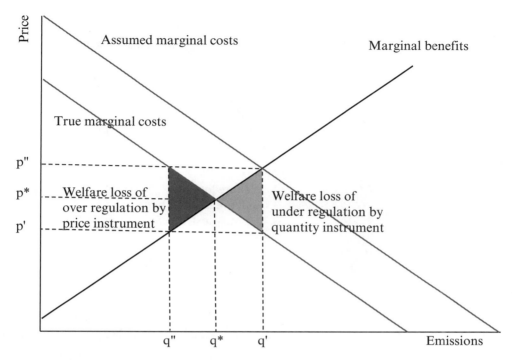

Figure 4.1 Welfare losses for price and quantity instruments if the regulator assumes abatement costs that are too high

regulator believes that the costs are higher than they are. However, if the government uses a price instrument, it overregulates: The emissions tax is higher than it should be because the regulator believes that the efficacy is lower than it is.

Figure 4.1 shows the associated welfare losses. The underregulation with a quantity instrument leads to a welfare loss for the environment and a welfare gain for the emitter; the latter is larger than the former. The net welfare loss is denoted in blue. The overregulation with a price instrument leads to a welfare loss for the emitter and a welfare gain for the environment; the latter is larger than the former. The net welfare loss is denoted in pink.

In Figure 4.1 the net welfare loss due to overregulation exactly equals the net welfare loss due to underregulation. (The distributional effects are, of course, different.) This is by construction. Figure 4.2 repeats the exercise, but with a steeper marginal benefit curve. The net welfare loss of underregulation falls, and the net welfare loss of overregulation rises. Figure 4.3 repeats the exercise once more, but with a shallower marginal benefit curve. The net welfare loss of underregulation rises, and the net welfare loss of overregulation falls.

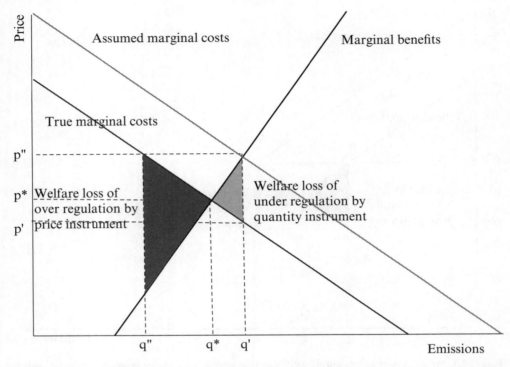

Figure 4.2 Welfare losses for price and quantity instruments if the regulator assumes abatement costs that are too high and the marginal benefit curve is steeper than the marginal cost curve

Figures 4.1, 4.2 and 4.3 illustrate one particular case, but the insight is much more general. If the marginal benefit cost curve is steeper (less steep) than the marginal abatement cost curve, then mistakes with quantity (price) instruments are more costly than mistakes with price (quantity) instruments. This is the Weitzman Theorem.

Climate change is driven by the stock of emissions. That implies that the marginal impacts of climate change do not change much if emissions are reduced or increased by a little. The effect of a change in emissions is dampened by the stock of emissions. In other words, the benefit cost curve is shallow. The marginal costs of emission reduction do, however, vary with emissions. Therefore, for a stock problem like climate change, mistakes with a price instrument (tax) are less costly than mistakes with a quantity instrument (tradable permits). The regulator should therefore levy a carbon tax, rather than create a market for emission permits.

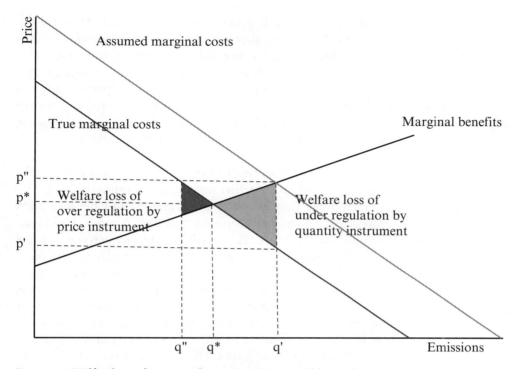

Figure 4.3 Welfare losses for price and quantity instruments if the regulator assumes abatement costs that are too high and the marginal benefit curve is shallower than the marginal cost curve

4.8 Initial allocation of permits**

The Weitzman Theorem fell on deaf ears. International tax harmonization is a political non-starter, even in the otherwise tightly integrated European Union. New taxes are anyway toxic in some jurisdictions. Tradable permits therefore play a substantial role in actual and planned climate policy around the world.

With tradable permits, as described above, the regulator sets an overall cap on emissions. The overall emissions cap is then split into units and each emitter receives a certain amount of permits to emit. If a company finds that it has too few permits, it may buy additional permits from a company that has too many. But how does the government get the newly created permits to the appropriate emitters?

There are many ways in which the initial allocation of emission permits can be implemented. I discuss the four basic ones.

Grandparenting (sometimes call grandfathering) of permits is by far the most popular choice. Permits are allocated, for free, on the basis of emissions in the recent past. This method is popular because it confirms the status quo. Large emitters are faced with a new regulation, but get a large amount of free permits in return. No money changes hands. Grandparenting is also unfair. Bad behaviour (large emissions) in the past is rewarded with a large allocation, while good behaviour (low emissions) is punished. Fast-growing companies are disadvantaged relative to slow-growing companies.

In 2012, the aviation industry became part of the EU ETS. Emission permits were grandparented based on the actual emissions, per airline, in 2005. The two big discount airlines, EasyJet and RyanAir, grew rapidly between 2005 and 2012, whereas the three big incumbents, British Airways, Lufthansa, and Air France, grew more slowly. The two discounters fly newer, more fuel efficient planes. They also have a higher load factor, that is, they pack more people in a plane. They fly point-to-point, avoiding energy-intensive take-offs and landings, and avoid congested airports. Per passenger kilometre, on the same route, the discounters thus emit less carbon dioxide than the incumbent airlines. Nonetheless, the initial allocation of permits is relatively generous towards the incumbents, and will thus lead to a transfer of wealth from the discounters to the incumbents. The formerly state-owned incumbents, of course, have a much closer relationship to the regulator.

Greenhouse gas emissions are externalities: unintended and uncompensated consequences of economic activity. The welfare loss of an externality follows from the fact that it is uncompensated. An alternative way to allocate emission permits is thus to give them to the victims. This would help to restore efficiency (see Section 4.2) and it adheres to a basic notion of fairness: If you want to emit carbon dioxide, you would have the buy the right to do so from someone who would be hurt by your act. It would be complicated to allocate emission permits in this manner. Victims would need to be identified and their relative damage estimated. Most of the victims of climate change are yet to be born, so their likely ancestors will need to be found. This allocation is also politically impractical. The majority of victims live in poor countries, while emissions are concentrated in rich countries. A large transfer of wealth would be the result. Some argue that this is desirable anyway, but it is a political non-starter.

Alternatively, emission permits may be allocated on a per capita basis. This corresponds to a basic notion of fairness: Everyone is treated the same. At a deeper level, it may not be fair at all. People in colder countries need more energy, and they live there because of choices their ancestors made, not knowing about climate change and climate policy. Disabled people often need more energy too. A per capita allocation of emission permits has a basis in international law too: The atmosphere is the common property of human-kind. In this view, the government has committed an injustice by grandpar-enting permits. It expropriated us, the people, and gave what is rightfully ours to private companies. A per capita allocation is unrealistic, however, because it would imply a large transfer of wealth.

One of the main issues with any free allocation of emission permits is that the market starts without a price. In the beginning, no one quite knows what permits are worth. The market is thin and erratic as a result. Therefore, as a fourth alternative way to initially allocate emission permits, an auction may be organized. Permits are sold to the highest bidder. Traders know the price. The regulator gains a substantial amount of revenue that it can use to lower taxes, to compensate victims, or to put in the president's bank account. In a distributional sense, auctioned permits are equivalent to taxes.

4.9   Initial and final allocation of permits*

Figure 4.4 depicts a barter trade between two people. One agent benefits from a polluting activity, and the other suffers from pollution. A social planner would set the level of pollution where the marginal benefits of the polluting activity equal the marginal costs of pollution.

Alternatively, the regulator may allocate explicit property rights to either party and organize a market. Suppose that the pollutee has the right to a pris-tine environment. The polluter then has to compensate the pollutee. As long as the marginal benefit of the polluting activity is greater than the marginal loss, the two parties should be able to strike a mutually advantageous deal. It is in both parties' best interest to agree on that level of pollution where the marginal costs equal the marginal benefits. This is illustrated in Figure 4.5.

Now suppose that the polluter has the right to pollute. The pollutee then has to compensate the polluter for emission reduction. As long as the marginal loss of reducing the polluting activity is greater than the marginal benefit of reduced pollution, the two parties should be able to strike a mutually advantageous deal. It is in both parties' best interest to agree on that level of pollution where the marginal costs equal the marginal benefits. This is illustrated in Figure 4.6.

Figures 4.5 and 4.6 illustrate the environmental interpretation of the Coase Theorem. More generally, the initial allocation of property rights does not affect the final allocation. Regardless of who gets the tradable emission

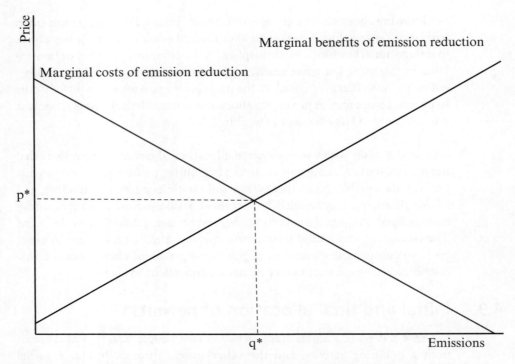

Figure 4.4 Marginal costs and benefits of emission reduction, optimal quantity and optimal price

permits, the market allocates them in the same way. The distributional consequences – equity – are different but independent of the market allocation – efficiency. The Coase Theorem separates equity and efficiency.

4.10 International trade in emission permits***

Long-distance trade precedes the formation of the nation state. Trade is the "natural" state of the economy. States erect artificial barriers to trade, but this typically slows down trade rather than end it.

The same holds for emission permits. If two countries each have a market in carbon dioxide emission permits, but with different prices, it would be mutually beneficial for one country to export permits and the other to import permits.

Emission permits are not goods or services, however. Permits are government licenses. A license by a foreign government is worthless unless it is explicitly recognized by the home government. The same is not true for, say,

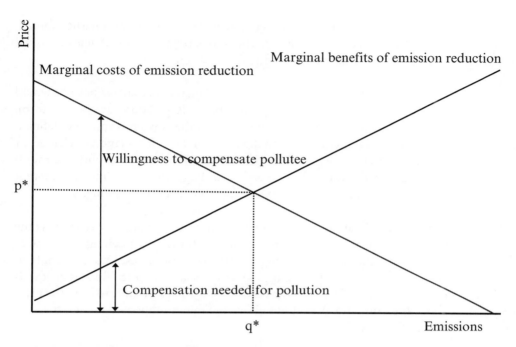

Figure 4.5 Marginal costs and benefits if there is a right to zero pollution

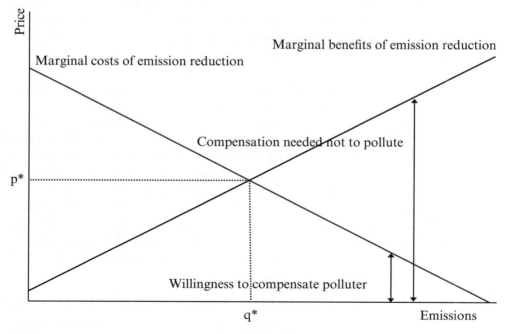

Figure 4.6 Marginal costs and benefits if there is a right to unlimited pollution

a barrel of oil extracted in a foreign country or a haircut by a barber abroad. International trade in emission permits thus requires explicit acts of mutual recognition by all states involved.

If regulations are uniform, the international market in tradable permits would be seamless and have a single permit price. Regulations are not uniform, however. Emission permit markets differ in the way emissions are defined. Does the market include carbon dioxide emissions from land use change, or non-CO_2 greenhouse gases? Markets differ in the way emissions are monitored, and in the way rules are enforced. Regulators may impose different penalties for infractions, or have an explicit price floor or ceiling.

Enforcement in an international emission permits market is as weak as the weakest national enforcement. The international price ceiling is as low as the lowest national price ceiling (if the regulator releases more permits to keep the price below the ceiling). The international price floor is as low as the lowest national floor (if the regulator buys back permits to keep the price above the floor; any price floor above the minimum would invite a game of beggar-thy-neighbour).

National regulators are not powerless, however, against importing weak regulations from abroad. First of all, unwanted countries can be excluded. Countries can also impose import tariffs or quotas, and differentiate these by country of origin. Emission permits are government licenses and therefore not subject to WTO rules.

Countries can also impose conversion rates. For instance, an imported permit may be set equal to, say, 90% of a domestic permit. That is, 1 tonne of imported carbon dioxide may be deemed equal to 900 kg of domestic CO_2. This conversion rate may reflect different definitions of emissions, or different enforcement. Care should be taken that these conversion rates do not create opportunities for carry trade. As conversion rates are set by government fiat and therefore cannot be arbitraged away, an international body should set an internally consistent set of rates. Alternatively, if countries announce conversion rates after the end of the trading and enforcement period, conversion rates would emerge in the permit market much like exchange rates emerge in the currency market, reflecting the beliefs of traders in what the final decision will be.

In sum, while national emission permit markets could be merged to form an international market, to the benefit of all countries involved, market regulation should be considered more carefully in this case.

4.11 Technological change**

The cost of greenhouse gas emission reduction is driven by the price difference between fossil fuel and carbon-neutral energy. Abatement costs would fall if technological change can be directed towards alternative energy sources. Similarly, abatement costs would fall if the rate of energy efficiency improvement can be accelerated. Furthermore, there is a large variation in energy efficiencies between similar activities in different companies, sectors, and countries. Emissions would fall if energy technology would diffuse faster.

Technological change should be treated with care. It would be great if technological change could be accelerated at zero cost. However, there may be large opportunity costs if technological progress is redirected towards energy. On a cost basis, energy is a few percent of the total economy. Costs would be substantial if we accelerate technological progress for energy at the expense of decelerating technological progress for the rest of the economy. In the short term, there is a fixed supply of smart and creative people. More people working on R&D in energy means fewer people working on R&D in ICT, medicine, and so on.

In the long term, the number of smart and creative people can be boosted, primarily by improving nutrition, health care, and education. The academic world is dominated by white men, not because they're smarter, but rather because they're privileged. Changing this is beyond the scope of climate policy, however.

Technological change is important. How can the government stimulate technological progress?

Technological progress comes in three stages: invention, innovation, and diffusion. Invention is a new idea or a new blueprint. Innovation takes an existing idea or blueprint and turns it into a product or service that can be sold, or a process that can be implemented. Diffusion takes the new product from its first sale to a substantial market penetration. All three stages are important, but they require a different set of skills and a different set of policy interventions (if any).

Invention requires smart people being creative. It cannot be forced, but it can be stimulated. Invention is primarily done in universities, research institutes, and some corporate laboratories. Inventions rarely generate intellectual property rights. Inventions instead contribute to the global stock of knowledge. Inventors are motivated by glory and curiosity rather than by money.

A country can best stimulate invention by rewarding its researchers for doing what they happen to be good at, and of course by supporting universities and research institutes.

As argued in Chapter 2, decarbonization of the economy does not require new inventions. We know how to supply the world's energy demand many times over without emitting carbon dioxide. The problem is that carbon-neutral energy sources are unproven, impractical, or expensive. Innovation and diffusion are therefore more important than invention.

Innovation is best done by corporate researchers. Innovation is not about creating something new, but rather about turning something that works in theory into something that works in practice. There are a few successful government innovations, and many unsuccessful ones. Corporate innovation has a high failure rate too, of course, but not nearly as bad as the public sector. Innovators take risks and are motivated by the prospect of making it big. The role of the government is to incentivize large companies to put its smartest people on climate-friendly research and development; and start-ups to focus on alternative energy.

Diffusion is best done by entrepreneurs. It is about getting ever more customers to buy the new product. Diffusion is about turning something that works into something that people want. Diffusers are motivated by the prospect of a steady and growing stream of profits. Government can help through regulation, but it is not a core task of the public sector to supply the market. The policy instruments to incentivize diffusion are discussed above: taxes and tradable permits.

In sum, climate-specific technology policy should focus on innovation rather than invention or diffusion. The government has a number of instruments to stimulate innovation.

Patents are probably the key instrument. Innovations can be copied by other companies. If the embedded knowledge cannot be kept a secret – for instance, because the product can be reverse-engineered – patents provide legal protection against copycats. Essentially, patents give a temporary monopoly on a particular technology. The patent holder either exclusively makes the product, or licenses other companies to do so. The monopoly rents reward the innovator for the effort made in innovation and the risks taken. If patents are properly designed, the efficiency loss due to monopolistic supply is smaller than the welfare gain from accelerated innovation.

Patents are a generic instrument that serves all innovation, not just climate-friendly innovation. Other widely used instruments are R&D subsidies and tax breaks. There are two problems with this. First, it rewards effort rather than success – and perhaps not even that as creative accounting may be just as effective in reaping these subsidies. Second, R&D subsidies are often very specific. This is, politicians or civil servants decide which particular technology is worthy of support. It may be that politicians and civil servants have excellent foresight into what products are likely to succeed in the market place, but more often than not government backs the wrong horse.

Economists refer to this as "picking winners". Governments are bad at picking winners. Politicians know a lot about politics, and civil servants know a lot about the civil service; they are less well-versed in the private sectors. Besides, the public sector does not bet its own money and thus lacks a key device to discipline risk taking.

The government is a large consumer. It can use its buying power to back particular products that are not quite ready for the market. Again, the government is picking winners. This strategy of selective procurement also means that running the public sector is more expensive than need be, and may imply that civil servants are saddled with experimental products with teething problems. Selective procurement is a bonanza for lobbyists.

Government procurement only works with products that are already on the market. The government may also opt for conditional procurement. Imagine, for instance, that there is a trust, supported by among others the Gates Foundation and the UK government that will buy 100 million doses of malaria vaccine from the first company that sells such vaccines for $1 or less. This would guarantee a market for a yet-to-be-invented product. This is not picking winners, because the government specifies the outcome rather than the technology. It rewards success rather than effort. The same model can be applied for climate policy.

Instead of promising to buy particular goods, the government can also forbid its substitutes. The Montreal Protocol on Substances that Deplete the Ozone Layer, for instance, forbade CFCs, thus creating a market for HFCs. The chemical industry in the Netherlands voluntarily agreed that every new installation be in the global top five with regard to energy efficiency. The government could consider banning the sale of cars that are more than X% less fuel-efficient than the best car on the market (within a class and price range). This instrument creates an incentive to innovate: the ability to force out the competition. It lets innovators, rather than civil servants and politicians, set the pace of innovation. It rewards success not effort. And it does not pick winners.

Guaranteed procurement or greater market share can be regarded as a "prize" for winning the technological race. The government can also grant actual prizes. This was a popular policy in the 18th and 19th centuries. More recently, the X Atari prizes have been reasonably successful. The prize for space flight, for instance, was $10 million but generated research worth $300 million; and may have kick-started a new market.

Innovation is best stimulated, however, by credible abatement policy. Innovation is an investment in the future. It is a bet that there will be a market for the product-to-be-invented. In the case of greenhouse gas emission reduction, the demand is primarily driven by government policy. If companies do not believe that there will be climate policy in the future, they will not innovate. From this perspective, a carbon tax is preferred. Taxes are rarely abolished, and tend to go up. Subsidies, on the other hand, are often short-lived. Permit prices go up and down. Direct regulation is also unpredictable, and there is no incentive to innovate beyond the target.

4.12 Emissions trade in practice: The EU Emissions Trading System**

The EU Emissions Trading System (or Scheme as it used to known; EU ETS) is the largest market for emission permits in the world, and the only multinational one. The EU ETS is now in its third phase. The first phase, 2005–2007, was primarily a test phase. The second phase, 2008–2012, helped Europe meet its commitments under the Kyoto Protocol. The current, third phase, 2013–2020, reflects the EU's unilateral commitments to control greenhouse gas emissions.

The EU ETS covers the emissions of 11 000 installations (not companies) in 31 countries. Australia plans to join by 2015. Some 45% of all greenhouse gas emissions fall under the EU ETS. Included are carbon dioxide from power and heat generation, metal production, pulp and paper, bulk chemicals, and mineral products; nitrous oxide from the production of acids; and perfluoro-carbons from aluminium production. Carbon dioxide from intra-Union aviation is also covered; coverage of extra-Union flights is suspended. There is a double selection. Besides the sectoral/gas criteria listed above, only installations that emit more than a threshold are included.

Permit markets can be created at any point in the production cycle. In an upstream market, emission permits would be needed for the exploitation and importation of fossil fuels. The problem is that there tend to be few companies in these sectors, so that market power is an issue. In a downstream market, emission permits would be needed for the consumption of goods and services in which carbon dioxide is embedded. This would be administratively costly as there are so many participants. Some people would find it difficult to manage their individual carbon account. The EU therefore opted for a mid-stream market.

Emission permits are held in electronic registries. Permits can be traded over-the-counter (that is, directly between two emitters), on a number of exchanges, and via brokers. Derivatives markets quickly appeared. More recently, permits can be bought at auctions too.

Emission permits are fungible within each phase. That is, emission permits for the third phase can be used at any time between 2013 and 2020. Between phase 2 and 3, emission permits can be banked but not borrowed. That is, a 2008–2012 emission permit is still valid after 2012. A 2013–2020 emission permit, on the other hand, was not valid before 2013.

Initially, permits were grandparented, that is, allocated for free to companies on the basis of past emissions. Over time, more and more permits are auctioned. In 2013, 40% of permits were auctioned. This should rise to 100% in 2020. These permit auctions are the second direct source of revenue for the European Commission.[2]

The EU ETS has had a number of teething problems, some of which could have been avoided. Initially, permits were allocated by the Member States. As the EU ETS covers only about half of the emissions, and constraints on the other half are not enforced, every Member State allocated more permits to its companies than it should have, in the hope of creating a new export industry. When the market collapsed under the oversupply, the European Commission took over the allocation of permits.

In the beginning, VAT treatment of permits was different between Member States. Carousel fraudsters bought permits in countries with no VAT and sold them in countries with VAT. Instead of transferring the VAT to the rightful treasury, the company was folded and the monies laundered. Several people ended in jail. Many more are probably lazing in the sun. Since 2010, VAT treatment has been harmonized.

There were administrative problems too. Several electronic registries were hacked, and trading had to be suspended. In Romania, the civil servant in charge of emissions monitoring went on maternity leave and was not replaced. Monitoring is essential for the integrity of the permits. The UNFCCC duly suspended Romania. Lithuania and Slovakia were also suspended for irregularities in emissions monitoring. The EU ETS did not follow suit.

Liability for emission permits is seller beware. Under buyer beware liability, the buyer is liable for the product after the sale is completed. Fruit is an example. As soon as you have paid for a rotten orange, it is your problem. Seller beware liability is the opposite. It is rare, and typically only applies to situations where information asymmetries are strong. In many jurisdictions, if you buy a second-hand car, you can return it if problems emerge within a certain period after the sale. If you buy a new-build house, the builder is liable for structural defects for a number of years.

Liability for emission permits is seller beware. If a company emits carbon dioxide without holding a permit, a fine will be imposed. If a company sells fraudulent permits – that is, it sells its permits but does not cut its emissions correspondingly – then the companies that buy such permits are immune. The purchasing companies should report the incident to their national regulators, who in turn should contact the regulator of the company that committed the fraud (and notify the European Commission). The latter regulator should then impose a fine.

Although this seems acceptable in theory, practice is different. Law abiding companies in strictly regulated Member States do not need to worry about purchasing fake permits. Enforcement in the EU is as weak as enforcement in the weakest Member State. Three of the Member States have been suspended for monitoring irregularities by the UNFCCC. Organized crime has penetrated the government of two Member States. One Member State had a convicted fraudster as its prime minister. Another Member State falsified its National Accounts. Two Member States routinely falsified their milk and olive statistics to maximize EU subsidies. There is no reason to believe that regulated emissions equal the number of emission permits.

Initially, the price of emission permits was high. See Figure 4.7. It collapsed, however, when the extent of initial overallocation became known. Later, the price picked up again. Since the start of the Great Recession, prices have gradually declined. The reason is twofold. First, lacklustre economic growth means that emissions are low (see Chapter 2). Second, people do not seem to believe that emissions will start growing again in the foreseeable future, or that emission targets will not be tightened.

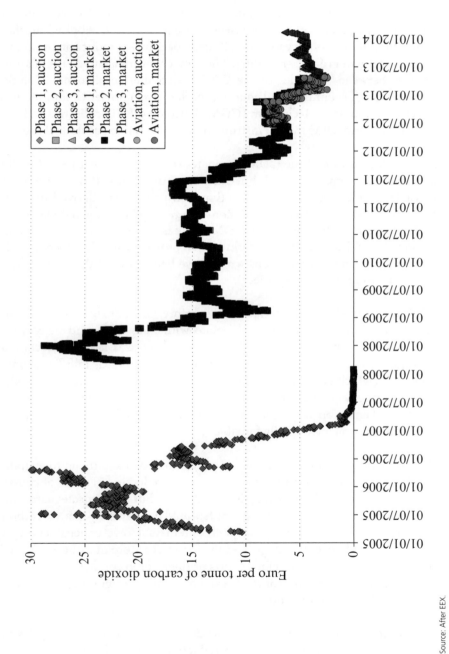

Figure 4.7 The price of greenhouse gas emission permits in the EU ETS

Source: After EEX.

4.13 Emissions trade in practice: The Clean Development Mechanism**

The Kyoto Protocol of the United Nations Framework Convention on Climate Change (UNFCCC) foresees in three "where flexibility" mechanisms, that allow a country to reduce its costs of emission reduction by investing in abatement in another country. These mechanisms are trade in emission permits between countries of the OECD, Activities Implemented Jointly between OECD countries and countries of the former Soviet Union, and the Clean Development Mechanism (CDM) between OECD countries and the rest of the world. Only the last instrument has been put to substantial use.

A key characteristic of the CDM originates from the fact that most OECD countries have emission targets but other countries do not. It is therefore not possible to trade emission permits. Emission permits are derived from an emission allocation, and total emissions can readily be compared with total emission permits. Instead, in the CDM, there is trade in Certified Emission Reductions (CERs). CERs are defined on a project basis. CERs are the difference between what emissions would have been without the project, and actual emissions (with the project).

CERs are defined by a counterfactual. Therefore, CERs are bureaucratic constructs. The bureaucracy is quite elaborate, with many forms and many committees. This implies that the CDM is skewed towards larger projects (so as to justify the fixed cost of project approval) and towards middle-income countries (which have the necessary expertise to get the project approved). This also implies that the price of CERs has always been a few euros below the price of emission permits in the EU ETS, even though the two certificates are legally equivalent and fully fungible.

Because CERs are counterfactual and project-based, it is difficult to guarantee that the certificates represent real emission reduction. That is of course one reason why the bureaucracy is so elaborate. But current safeguards are insufficient. For instance, planting forests would permanently remove carbon dioxide from the atmosphere – if the trees remain. However, legal tenure of the land – and hence the integrity of the forest as a store of carbon – means little in countries with capricious courts or rulers, or in countries where contracts are badly enforced. Contracts can be annulled or ignored, leaving the holders of CERs empty-handed.

As another example, a project to close down a factory for motorcycles in Indonesia would qualify as emission reduction and be eligible for CERs –

even if this project leaves the demand for motorcycles unchanged and hence emissions from motorcycle production. Such a project would be profitable if the revenue of the sale of CERs is greater than the cost of buying and winding down the business.

As a third example, some industrial gases, that are very potent greenhouse gases, can be made cheaply. Projects that closed the factories of such gases used to qualify for CERs. This was discontinued when it became apparent that some factories were built with the sole purpose of closing them down and selling the resulting CERs – while selling on the manufacturing equipment to the next such scam.

Despite these problems, 6000 CDM projects have now been approved. There clearly is a demand for certificates that can be held in lieu of emission reduction at home.

 FURTHER READING

There are many books on specific aspects of (international) climate policy, particularly on emissions trading. A good overview is Thomas Sterner and Jessica Coria's *Policy Instruments for Environmental and Natural Resource Management* (2002).
IDEAS/RePEc has a curated bibliography on this topic at http://biblio.repec.org/entry/tdc.html.

 EXERCISES

4.1 In Section 4.8, Weitzman's Theorem on prices versus quantities is illustrated with the case in which the regulator assumes that emission abatement is more expensive at the margin than it really is. Repeat the exercise assuming that the regulator believes that marginal abatement costs are lower than they really are. Repeat the exercise again assuming that the regulator believes that marginal damages costs are lower than they really are.

4.2 Consider I companies with emission reduction costs $C_i = \frac{1}{2}\alpha_i R_i^2$. Companies have baseline emissions E_i and an emissions target T_i. Without permit trade, companies have to cut emissions such that $T_i = (1-R_i) E_i$. What are emission cuts with permit trade? (Hint: Assume that all companies are price-takers.) What is the permit price? Assume that $I = 3$, $\alpha_1 = 1$, $\alpha_2 = 2$, $\alpha_3 = 3$, $E_i = 100$ and $T_i = 90$. What is the difference in costs with and without trade? What happens to the permit price and emission reductions if $T_1 = 80$ and $T_3 = 100$?

4.3 Read and discuss:

- **D. Cameron, N. Clegg and N. Huhne (2011), *The carbon plan: Delivering our low carbon future*, London: HM Government. https://www.gov.uk/government/publications/the-carbon-plan-reducing-greenhouse-gas-emissions-2.
- **P. Morell (2007), 'An evaluation of possible EU air transport emissions trading scheme allocation methods', *Energy Policy*, **35**, 5562–5570.
- **J.D. Fitz Gerald and R.S.J. Tol (2007), 'Airline emissions of carbon dioxide in the European trading system', *CESifo Forum*, 1/2007.

- ***C. Boehringer, H. Koschel and U. Moslener (2008), 'Efficiency losses from overlapping regulation of EU carbon emissions', *Journal of Regulatory Economics*, **33** (3), 299–317.
- ***A. Anger and J. Koehler (2010), 'Including aviation emissions in the EU ETS: Much ado about nothing? A review', *Transport Policy*, **17**, 38–46.
- ****T. Requate and W. Unold (2003), 'Environmental policy incentives to adopt advanced technology: Will the true ranking please stand up?', *European Economic Review*, **47**, 125–146.
- ****C. Fischer, I.W.H. Parry and W.A. Pizer (2003), 'Instrument choice for environmental protection when technological innovation is endogenous', *Journal of Environmental Economics and Management*, **45**, 523–545.
- ****C. Fischer and R.G. Newell (2008), 'Environmental and technology policies for climate mitigation', *Journal of Environmental Economics and Management*, **55**, 142–162.

NOTES

1 Note that the compensation flows in the opposite direction if the externality is positive.
2 Fines for violations of competition law are the other direct source of funding to the EU.

5
Impacts and valuation

TWEET BOOK

- Climate change affects managed and unmanaged ecosystems. Specialist and marginalized species will be hit hardest. #climateeconomics
- Global food production will first increase, mainly due to CO_2 fertilization, but decrease later in the century. #climateeconomics
- Climate change increases the demand for drinking, cooling and irrigation water. Both floods and droughts get worse. #climateeconomics
- Energy demand will go down in winter, up in summer. Labour productivity will decline, unless air-conditioned. #climateeconomics
- Cold-related deaths will go down, heat-related ones up. Infectious diseases, like malaria and diarrhoea, will increase. #climateeconomics
- Sea level rise will cause land loss, wetland loss, floods, saltwater intrusion; and require costly protection measures. #climateeconomics
- Adaptation substantially reduces the negative impacts of climate, and may even change their sign. #climateeconomics
- The impacts of climate change are many and diverse. A superindicator is needed to assess its seriousness. #climateeconomics
- Money was invented to compare and add the value of diverse goods and services, and indeed income. #climateeconomics
- Behaviour in related markets (housing, recreation, labour) can be used to estimate money value of environmental goods. #climateeconomics
- Revealed preference methods only reveal the direct consumption value of the environment. #climateeconomics
- Stated preference methods can reveal any value, but people do not necessarily speak the truth. #climateeconomics
- Measured values are multifaceted, difficult to generalize and thus hard to extrapolate to future climate change. #climateeconomics
- Willingness to accept compensation is (much) larger than willingness to pay because of loss aversion and imposed risk. #climateeconomics

5.1   Impacts of climate change**

The impacts of climate change are many and diverse.

Changes in temperature, rainfall, cloud cover, wind direction, wind speed, and so on, would directly affect plants and animals, and those effects would have further impacts through predation, competition and other ecological interactions. This is true for both managed and unmanaged ecosystems. Some of these impacts will be positive, and others negative. Some impacts will be small, and others large.

The biggest impacts will be seen for marginalized species and specialists. Marginalized species, by definition, are at the edge of survival. Any change, including climate change, could either push them over the edge to extinction or dramatically expand their ecological niche.

Specialists, by definition, thrive under very particular conditions. By contrast, generalists can live most anywhere. Often, specialists do not thrive – but rather survive where others cannot. If climate would change, their ecological niche would disappear. Although it might well re-appear elsewhere, it is doubtful whether the new and the old niches are sufficiently connected to allow for migration. This is easily illustrated with Edelweiss, a pretty little plant that lives where few others can, high up in the Alps. If the world would warm, Edelweiss would have to move north – but it cannot jump from mountain top to mountain top, it cannot compete with the plants that live in the valleys, and there are no mountains immediately north of the Alps.

Climate change would not mean that large parts of the planet would turn into a lifeless moonscape. Rather, nature would become duller, with fewer species covering larger areas.

Online Figure 5.1 illustrates the scale at which this might occur. Coloured areas denote a wholesale change in the composition of the ecosystem. The details depend on the model and the climate scenario, but the scale does not.

Agriculture and forestry would also be affected by climate change. Most crops, but weeds too, would grow faster because of the higher concentration of carbon dioxide in the atmosphere. This is a fertilizer, and allows plants to manage their water more efficiently. Some crops would benefit from warmer and wetter conditions. Other crops may suffer drier conditions or be less heat tolerant. The net impact depends on the crop and its location.

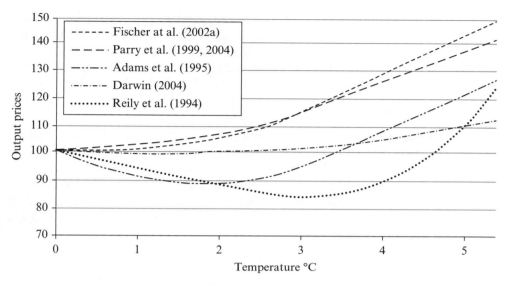

Source: IPCC WG2 AR4 Chapter 5 (references in the figure are detailed in this chapter).

Figure 5.1 The impact of climate change on food prices

Figure 5.1 shows the aggregate impact according to five different models. The indicator is the world market price for food. In the next few decades, world food production may well expand because of climate change, suppressing food prices. In the longer run, however, climate change is likely to reduce food production, pushing up food prices.

Climate change affects water resources, directly through precipitation and evaporation and indirectly through changes in water use. This would have an impact on agriculture, nature, drinking water, and inland navigation. It would also affect power generation, which often uses water as a coolant. Less or hotter water would constrain that. There are further impacts on energy supply. Wind and wind power and cloud cover and solar power come immediately to mind, but thermal plants are less efficient when it is hot, and resistance increases with the temperature of transmission cables. Energy use would be affected too. Demand for cooling energy would increase, and demand for heating energy would fall. Construction and transport are interrupted by weather events such as cold spells, heat waves, floods, and fog. Every winter, tourists flock to mountains to ski while beaches are popular in summer. Climate change would affect the attractiveness of holidays in particular locations.

Sea level rise would have a number of effects. Coastal erosion would increase, and floods would be more frequent or intense. Saltwater would intrude into groundwater. Many fear that sea level rise would lead to the disappearance of atoll islands, which often do not reach more than a metre above the current high sea level. Saltwater intrusion is likely to make many of these islands inhabitable decades before they finally disappear beneath the waves. Coastal wetlands may drown, particularly if coastal defences prevent inland migration.

Human adaptation is critically important in all impacts of climate change, but perhaps best illustrated for sea level rise. Coastal defences were known to the Ancient Greeks and Chinese. As sea level rises, people will not sit on their hands while their buildings, roads and land are swept away. Dikes would be raised, groins built and beaches nourished. In many places, the cost of adaptation would be the main impact of climate change, and residual impacts would be relatively small.

Climate change would also affect labour productivity. The human body is in a thermal equilibrium with its environment. As any warm-blooded animal, we need to keep our body at a particular temperature. Shivering to keep warm and sweating to keep cool cost energy. Work raises our body's temperature. Sweating is less effective in humid conditions. So the human body is less able to do work when it is hot and humid. The productivity of outdoor, physical labour would fall if climate changes.

Through the same process, heat waves affect human health. Healthy people tire when it is hot. The bodies of the very young, the very old, and people with cardiovascular or respiratory disorder may give up altogether. Cold kills too. Like heat, cold creates physiological stress. During cold weather, people group together indoors, giving free reign to infectious diseases. Climate change would further affect human health through nutrition (see the discussion on agriculture above), through air pollution, and through vector-borne diseases such as diarrhoea, malaria and cholera.

5.2   Purpose of valuation*

Monetary valuation seeks to estimate the value of environmental goods or services that are not traded on markets. Market goods and services are routinely valued and property is frequently valued too, typically in preparation for a sale. The purpose of environmental valuation is different.

There are many impacts of climate change, some positive, some negative, some big, some small. Impacts vary over space and over time. The question whether or not climate change is a problem, and whether it is a big problem or small problem cannot be answered without aggregating the impacts. Monetary valuation serves this purpose. It puts all impacts in a common metric, money in this case, which is a prerequisite for aggregation.

Expressing the total and marginal impact of climate change in monetary terms is handy because it allows for an immediate comparison with the impacts of greenhouse gas emission reduction. It also allows for a comparison with other issues, and to the Gross Domestic Product. Furthermore, if the victims of climate change are to be compensated, it will likely be in the form of money.

Some people object to environmental valuation, or find it hard to understand how putting a price tag on something valuable is feasible or meaningful. Yet, money was invented exactly for this purpose. In a barter economy with N goods, there are $0.5*N*(N-1)$ prices. In a money economy, there are only N prices. That is probably why money was invented: To reduce the transaction and information costs of trade. And through the medium of money, some strange trade-offs are made. Working within a tight budget, students have to choose between a new pair of jeans, a night out, or a textbook. Professors can afford all those things, but have to make a choice between a boat, an extension to the house, or sending the kids to Harvard. Those things are incomparable at first glance, yet choices are made every day. Presumably, people can compare such things. Some things are worth the money they cost, and other things are not.

Monetary valuation of environmental goods and services is done for the purpose of improving decisions and making choices that are consistent with other choices we make. We would maximize environmental quality if that were costless. It is not. Sacrifices need to be made to reduce emissions. Some of these sacrifices are worth it, and others are not. Monetary valuation informs that decision.

5.3   Valuation methods: Revealed preferences*

There are a number of methods, and many variants, to value environmental goods and services. The more reliable but narrower ones use the actual behaviour of people and households. The travel cost method is the oldest method, and perhaps the most intuitive one. It belongs to the broader class of household production methods.

Consider your local park. If you would ask the visitors where they are from, you would learn that most of them come from the neighbourhood. Many live a block away and are in the park with their dogs or children. Some cycled or drove 10–15 minutes. Few have travelled across the country, and none across the world to be in your local park. That makes perfect sense. Your park is nice, but nothing special. There are many similar parks elsewhere. Why would anyone travel just to visit your park?

Now consider the Great Barrier Reef. There are many visitors. There are locals, of course, but relatively few. People fly all the way across the world to visit the Great Barrier Reef. Why? Because it is unique and spectacular!

The food that you buy is worth at least as much to you as the money you spent on that food. The movie that you see in the cinema is worth at least as much as the ticket you need to get in. You do not pay an entrance fee to get into your local park. However, you do spend time to get there, and you may spend money on a bus fare or something similar.

If you extend your visitor survey and ask people how long they needed to get to the park and how much money they spent getting there, you would find that many paid little, few paid more, and none paid a whole lot. You would find something that looks remarkably like a demand curve: Low price, high demand; high price, low demand. In fact, you have found a demand curve. If you integrate under the curve, you estimate the consumer surplus generated by your local park. If you then repeat the exercise for the Great Barrier Reef, you find that demand is still high at a high price.

Although conceptually clear, the travel cost method is beset with practical difficulties. Travel time is valuable, but how valuable exactly? In a perfect labour market, the wage equals the marginal value of leisure – but labour markets are distorted in many ways. Trips often serve multiple purposes (e.g., going to the park and the shop; visiting Sydney and the Great Barrier Reef) and that means that the travel cost needs to be apportioned to these purposes. Sometimes the trip is a cost (e.g., travelling alone in a hot and crowded train), and sometimes the trip is part of fun (e.g., travelling in cool and quiet train with friends). These problems can be overcome with a sufficiently detailed survey, plenty of data, and clever econometrics.

The second class of revealed preference methods analyzes household consumption. Hedonic pricing is the best known example. A house that sits in a beautiful environment is worth more than the exact same house that sits in an ugly environment. The price difference is an indication of the value of environmental beauty.

Like the travel cost method, hedonic pricing is conceptually straight-forward but difficult in practice. Builders are not stupid. They put the prettiest houses in the prettiest environments, and more ordinary houses elsewhere. Expensive houses attract more well-to-do home owners, who tend to be better educated and socially more attractive as neighbours. Such neighbourhoods tend to have better schools and other facilities. At the larger scale, wages compensate both for the local cost of living and for the attractiveness of the environment. In sum, the housing market is influenced by many things, and you need a large amount of observations and clever econometric methods to isolate the effect of the environment – but it can be done.

5.4   Valuation methods: Stated preferences*

Revealed preference methods have the advantage that actual decisions are analyzed. The disadvantage is that it considers only those values that are expressed, indirectly, in market transactions.

I care about whales. Nothing in my behaviour of the last 15 years has revealed that I do. I do not contribute to Greenpeace, the major NGO that campaigns for the preservation of whales, because I do not agree with their energy and climate policies. I do not go whale watching. I did that once. The whales did not show up. I will not do it again. You could have followed me around for 15 years, checked all my bank statements, and you would not have learned that I care about whales. Yet, I do. To find out, you need to attend my class or read my book. Or you could ask me.

Stated preference methods do exactly that. The contingent valuation method is the oldest and most widely used. It uses surveys (face-to-face, by phone, over the Internet) that include questions such as "how much would you be willing to contribute to help preserve the population of grey-blue humpback whales in the North Atlantic?" Researchers have now moved away from open-ended questions (as above) to single-bounded – "would you be willing to pay more than £50 per year" – or double-bounded – "would you be willing to pay between £50 and £75 per year" – questions plus randomization of interviewees. More recently, contingent choice methods have become popular. Here, interviewees are asked to choose between sets of attributes – "would you rather contribute £50/year and have a population of 8,000 humpback whales or contribute £75/year and have a population of 10,000 humpback whales?" Contingent choice has gained acceptance because it resembles other purchase decisions more closely, and because it reveals more about the environmental characteristics that interviewees care about.

The main advantage of stated preference methods is that you can value anything: consumption of environmental services (as in revealed preferences), option values (I am not using it now but I may want to use it later), bequest values (I do not care much but I would like my children to enjoy it), and existence values (I am happier because I know that there are whales out there).

The main disadvantage of stated preference methods is that interviewees do not put their money where their mouth is. Interviewees may therefore take less care in expressing their preferences, or they may try and mislead the interviewer.

Interviewees may be influenced by the interviewer; they may be repelled (and thus give a low value or none at all) or may try to impress (and thus give a high value). Interviewees may realise that they are not being asked to contribute, but rather that they are asked about spending government money, and that their answer will be averaged with other interviewees. Interviewees may care about the environmental service in question, but object to the suggested way of delivery (e.g., higher taxes for environmental protection).

Standard micro-economic theory assumes that people are rational and fully informed. In fact, experience is a better description. If people make routine decisions in a familiar environment – for example, the weekly trip to the supermarket – they buy the stuff that they want, and pay a reasonable price for an acceptable quality. If you let people make the same decisions in an unfamiliar environment – a foreign supermarket, say – errors creep in. For decisions that are not routine, people gather information by searching the Internet and talking to friends and family before making a choice.

Contingent valuation and contingent choice methods put interviewees in a situation that is unfamiliar and asks them to make a decision that is not routine. The results are therefore noisy.

Stated preference methods are now implemented with a standard battery of tests that check and correct for the many biases that may creep into the results. Although these methods are applied routinely in public policy making and litigation, it is a very active research field and results are less reliable than we would like them to be.

5.5 Issues for climate change: Benefit transfer**

There are two problems with valuation methods that are particularly relevant for the impact of climate change.

Primary valuation is expensive. As suggested above, conceptually straight-forward ideas are difficult to put into practice. An applied revealed preference study easily employs someone for a full year. Stated preference studies are considerably more time-consuming. Therefore, there are only a finite number of primary estimates. Because the methods are under continuous development, researchers often go back and re-value a good or service that was previously studied.

Case studies are great for science and replication is better, but in order to inform public policy we need a comprehensive coverage of all goods and services affected in every location. Extrapolation is required.

Two techniques are used to extrapolate primary estimates from one location to another, from one issue to another, and from one time to another. First, existing estimates are subject to a meta-analysis, a set of statistical techniques to discover empirical regularities in previously published results. This may reveal, for instance, that richer people are willing to pay more for environmental protection; an income elasticity is estimated. Second, the estimated relationships are used to extrapolate from the observed sites to all sites of interest. This is known as benefit transfer (by transfer function).[1]

This is a reasonable approach. Unfortunately, meta-analytic regressions have low explanatory power, and tests of the validity of benefit transfer show large errors. The reason is twofold. The data are noisy, and values are highly context specific. Idiosyncrasy cannot be predicted.

This matters for climate change because the relevant impacts occur in the future, which cannot be observed. Furthermore, there have been few primary studies to value the impacts of climate change, so benefits are transferred from other issues, such as occupational health, air pollution, and eutrophication. Finally, valuation studies have disproportionally focused on rich countries, but the impacts of climate change are concentrated in poor countries. Valuation of the impacts of climate change is thus a particularly uncertain business.

5.6   Issues for climate change: WTP versus WTAC**

The other problem with valuation that is especially important for climate change is as follows. Above, I conceptualize the question as the willingness to pay (WTP) to acquire an environmental good or improve an environmental service. You may also conceptualize the question as the willingness to accept compensation (WTAC) for a deterioration of environmental quality.

Consider the following example. Someone knocks on your door, tells about the plans to convert the parking lot down the road into a park, and asks you for a financial contribution. Now contrast this to the situation where someone knocks on your door, tells about the plans to convert the park down the road into a parking lot, and offers you financial compensation.

Objectively, the comparison is the same: tarmac and cars versus trees and grass. There are differences, though. In one case, you are asked to contribute and are therefore constrained by your budget. In the other case, you are offered money and are therefore constrained by your perception of the other party's budget. Under standard micro-economic assumptions, the budget constraint makes a small difference. The difference between WTP and WTAC can be large, however, if income is a poor substitute for the good being valued.

There is another difference: You may be emotionally attached to the existing park because of the happy memories you have of the place. You may have often walked your dog there, or got your first kiss. You cannot be emotionally attached to a hypothetical park that was introduced to you a few minutes before. The amount you would be willing to accept as compensation for the loss of a park is thus greater than the amount you would be willing to pay to acquire a park.[2]

Empirical studies indeed show this. See Figure 5.2. The willingness to accept compensation for the loss of a good or service is larger, often a lot larger, than the willingness to pay for the same good or service. Four explanations have been offered. The budget constraint and emotional attachment are two.

Studies have shown that people are loss averse. They attach a value to the status quo. Losing something is worse than gaining the same thing is good. Loss aversion has been shown to occur even for routine, low-worth goods (e.g., coffee mugs) that were acquired less than an hour ago. If you give students a mug and try to buy it back from them, they demand a price that is much higher than they would pay for the same mug in the shop next door. It is easy to describe such behaviour as irrational. It creates a dilemma for public policy. Do you educate people to be more rational, ignore this aspect of people's preference, or seek to reflect the strange will of the people in the government's decisions?

The fourth explanation of the difference between willingness to accept compensation and willingness to pay is that voluntary risks are viewed differently than involuntary risks. Suppose that you get drunk, go joy-riding, get into an accident, and lose a leg. That would feel bad. Now suppose

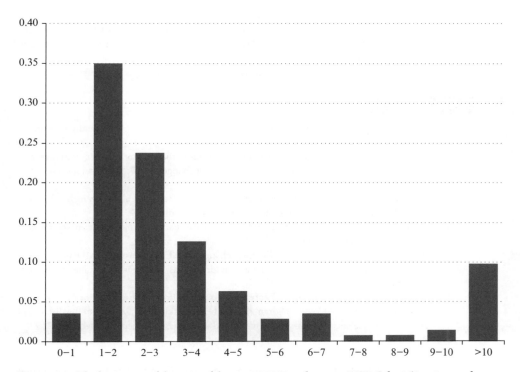

Figure 5.2 The histogram of the ratio of the mean WTP to the mean WTAC for 168 estimates from 37 studies

someone else gets drunk, goes joy-riding, gets you into an accident, and you lose a leg. That would feel worse – even though there is no objective difference: Your leg is gone. Context matters for valuation because people are social animals.

This matters for climate change. Do we formulate climate policy as us buying a better climate for our children, or do we conceptualize the problem as us imposing a worse climate on our children and offering them compensation in return? Do we formulate climate policy as rich people buying a better climate for their richer children? Or as rich people imposing a worse climate on the children of the poor? Do we view carbon dioxide emissions as necessary for survival? Or as an indulgence of a luxurious life style? The value of climate change impacts would be different, depending on how the question is framed.

 **FURTHER READING**

Valuation methods are part of any textbook on environmental economics. A good introduction is Garrod and Willes' *Economic Valuation of the Environment: Methods and Case Studies* (2000)

but Braden and Kolstad's *Measuring the Demand for Environmental Quality* (1991) sets the technical standard. Daniel Kahnemann's *Thinking, Fast and Slow* (2011) is an easily accessible entry into some of the above material.

IDEAS/RePEc has a curated bibliography on this topic at http://biblio.repec.org/entry/tbb.html.

 EXERCISES

5.1 Climate change would change landscapes as vegetation responds. How could you estimate the value of changes in the landscape?

5.2 Climate change would affect human health through changes in weather extremes and vector ecology. How could you estimate the value of changes in risk to human mortality and morbidity?

5.3 Climate change would affect species abundance and may lead to local and even global extinctions. How could you estimate the value of changes in biodiversity?

5.4 Estimates of the value of the impact of future climate change are necessarily based on data from the present and past. How could you estimate future values?

5.5 Would the valuation of the impact of climate change be different if we phrase the policy question as "buying a better climate for our grandchildren" or as "compensating our grandchildren for climate change"?

5.6 Section 4.10 discusses the Coase Theorem and its application to the initial allocation of emission permits. Does the Coase Theorem need to be reconsidered in the light of the discussion on the difference between willingness to pay and willingness to accept compensation?

5.7 Read and discuss:

- **L.M. Brander, P. van Beukering and H.S.J. Cesar (2007), 'The recreational value of coral reefs: A meta-analysis', *Ecological Economics*, **63**, 209–218.
- **W.K. Viscusi and J.E. Aldy (2003), 'The value of a statistical life: A critical review of market estimates throughout the world', *The Journal of Risk and Uncertainty*, **27** (1), 5–76.
- ***J.K. Horowitz and K.E. McConnell (2002), 'A review of WTA/WTP studies', *Journal of Environmental Economics and Management*, **44**, 427–447.
- ***R. Brouwer (2000), 'Environmental value transfer: State of the art and future prospects', *Ecological Economics*, **32**, 137–152.
- ****R.O. Mendelsohn, W.D. Nordhaus and D. Shaw (1994), 'The impact of climate change on agriculture: A Ricardian analysis', *American Economic Review*, **84**, 753–771.
- ****K. Rehdanz and D.J. Maddison (2005), 'Climate and happiness', *Ecological Economics*, **52**, 111–125.
- ****O. Deschenes and M. Greenstone (2011), 'Climate change, mortality, and adaptation: Evidence from annual fluctuations in weather in the US', *American Economic Journal: Applied Economics*, **3**, 152–185.

NOTES

1   In olden times, researchers used values estimated for one location, issue and time and applied them, unadjusted, to other locations, issues and times. That is wrong.
2   Some people are emotionally attached to parking lots.

6

Impacts of climate change

TWEET BOOK

- Our best estimate is that global warming of 2.5K would make the average person feels as if she'd lost 0.9% of income. #climateeconomics
- There are only 17 estimates of the total economic impact of climate change, and our confidence is thus low. #climateeconomics
- Climate change is initially beneficial but these are sunken gains. Net impacts turn negative around 2.1K global warming. #climateeconomics
- Climate change will not stop at 3K warming, but impact studies stop there. Beyond that, there is speculation. #climateeconomics
- Poor countries are more vulnerable to climate change because they tend to be hotter and closer to biophysical limits. #climateeconomics
- Poor countries are more vulnerable because a larger share of their economic activity is exposed to the weather. #climateeconomics
- Poor countries are more vulnerable because they lack the means, the wherewithal, and the political will to adapt. #climateeconomics
- As poverty implies vulnerability, economic growth is an, often superior, way to reduce the impact of climate change. #climateeconomics
- The social cost of carbon is the benefit of reducing greenhouse gas emissions by a single tonne. #climateeconomics
- The social cost of carbon is an estimate of the desirable intensity of climate policy. #climateeconomics
- The social cost of carbon depends on many things, so there are many, widely different estimates. #climateeconomics
- The social cost of carbon is higher if the discount rate is lower, and its right tail is much fatter. #climateeconomics

6.1   Reasons for concern**

The impacts of climate change are many and diverse. The question whether climate change is beneficial or detrimental, big or large, depends on sector, location and time. Reading through Section 5.2, let alone the wider literature on which it is based, leaves one confused. Aggregate indicators are needed to

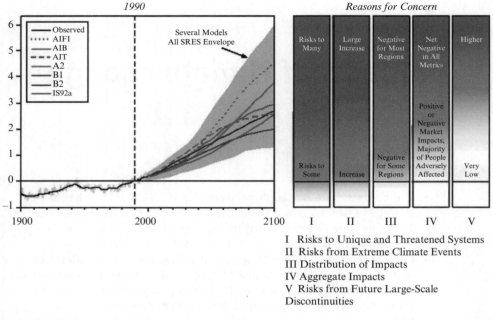

I Risks to Unique and Threatened Systems
II Risks from Extreme Climate Events
III Distribution of Impacts
IV Aggregate Impacts
V Risks from Future Large-Scale
Discontinuities

Source: IPCC WG2 AR3.

Figure 6.1 Projected climate change (left panel) and alternative reasons for concern about climate change

assess whether climate change is, on balance, a good thing or a bad thing, and whether it is small or large relative to the many other problems that we have.

Figure 6.1 uses alternative high-level indicators and displays them against projected climate change. The indicators are alternatives in that they would appeal to people with different attitudes.

Some people worry about the unique and the vulnerable, such as atoll islands or butterflies. If you are so inclined, climate change is a big concern. Local extinctions of butterfly species have been documented with climate change as the likely cause. Many butterflies have difficulty crossing open spaces and thus cannot migrate. If the limited climate change of the past century already caused such problems, the more rapid climate change of this century must be disastrous (from this perspective). Stringent climate policy is thus justified.

Other people only care about systemic impacts of climate change, such as changes in ocean currents and the melting of the polar ice caps. If you are so inclined, then climate policy can wait. The probability of these scenarios is minute, and it is not known how greenhouse gas emission reduction would affect those probabilities – indeed whether they would go up or down.

Yet other people may care about the impact of climate change on total economic welfare, or about the distribution of that welfare. Those concerns are discussed below.

6.2   Total economic impacts**

Figure 6.2 shows the 21 published estimates of the total economic impact of climate change. The numbers should be read as follows: A global warming of 2.5°C would make the average person feels as if she had lost 1.1% of her income. (1.1% is the average of the 9 dots at 2.5°C.)

Most of these estimates were derived as follows. Researchers used models to estimate the many impacts of climate change for all parts of the world, estimated the values of these impacts (using either market prices or the methods described in Chapter 5), multiplied the quantities and prices, and added everything up. This is the so-called enumerative method.

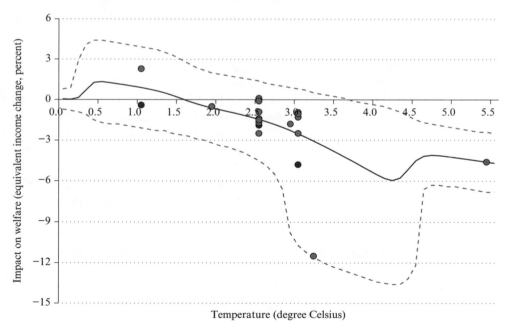

Figure 6.2 The global total annual impact of climate change expressed in welfare-equivalent income change

Other estimates involve regressions of some sort of a welfare measure on climate. Agricultural land prices, for instance, reflect the productivity of the land and hence the value of the climate that allows plants to grow. The main advantage of the statistical method is that is based on actual behaviour (rather than modelled behaviour as in the enumerative method). The main disadvantage is that climate *variations over space* are used to derive the impact of climate *change over time*.

Yet other estimates elicit the views of supposed experts, or use physical impact estimates to shock a computable general equilibrium model and derive a welfare estimate that takes all market interactions into account.

Figure 6.2 contains many messages. There are only 21 estimates, a rather thin basis for any conclusion. Statements that climate change is the biggest (environmental) problem of humankind are simply unfounded (that is to say, we do not know whether it is true or not).

The 9 estimates for 2.5°C show that researchers disagree on the sign of the net impact. Climate change may lead to a welfare gain or loss. At the same time, researchers agree on the order of magnitude. The welfare change caused by climate change is equivalent to the welfare change caused by an income change of a few percent. That is, a century of climate change is about as good/bad as a year of economic growth.

Considering all 21 estimates, it is suggested that initial warming is positive on net, while further warming would lead to net damages. This does not imply that greenhouse gas emissions should be subsidized. In Figure 6.2 the total impacts turn negative somewhere between 1 and 2°C. More importantly, the incremental impacts turn negative before that, around 0.5°C global warming. Because of the slow workings of the climate system and the large inertia in the energy sector, a warming of 2°C can probably not be avoided and a warming of 1°C can certainly not be avoided. That is, the initial net benefits of climate change are sunken benefits. We will reap these benefits no matter what we do to our emissions.

The uncertainty is rather large, however. The error bars in Figure 6.2 depict the 90% confidence interval. This is probably an underestimate of the true uncertainty, as experts tend to be overconfident and as the 21 estimates were derived by a group of researchers who know each other well.

The uncertainty is right-skewed. Negative surprises are more likely than positive surprises of similar magnitude. This is true for the greenhouse gas emissions: It is easier to imagine a world that burns a lot of coal than a world that rapidly switches to wind and solar power. It is true for climate itself: Feedbacks that accelerate climate change are more likely than feedbacks that dampen warming. The impacts of climate change are more than linear: If climate change doubles, its impacts more than double. Many have painted dismal scenarios of climate change, but no one has credibly suggested that climate change will make us all blissfully happy. In that light, the above conclusion needs to be rephrased: A century of climate change is no worse than losing a decade of economic growth.

The right extreme of Figure 6.2 is interesting too. At 3.0°C of warming, impacts are negative, deteriorating, and accelerating. It is likely that the world will warm beyond 3.0°C. Yet, beyond that point, there are few estimates only. There is extrapolation and speculation.

Figure 6.2 shows the world average impact for 21 studies. Figure 6.3 also shows the world average impact, but now for a single study. Impacts are shown as a function of time rather than warming. The shape is very similar, though: Initially, net impacts are positive, but they turn negative in the second half of the 21st century. Figure 6.3 also shows the impacts in the best off country and the worst off (continental) country.[1] As always, the average hides the extremes. The best off country is Canada (in this particular model, parameterization, and scenario), which is rich, well-organized and rather cold. The impact is positive throughout the 21st century, as are incremental impacts so that there is no incentive to reduce emissions. The worst off country is Guinea-Bissau (in this particular model, parameterization, and scenario), which is poor, chaotic, low-lying, disease-ridden and rather hot. Impacts are negative now already, improve a bit with economic growth, and then progressively get worse over the century.

Figure 6.4 shows results from the same model, but now for all countries in the year 2050. Countries are ranked from low to high per capita income (in 1995). In Figure 6.3, the world total impact is roughly zero. In Figure 6.4, the majority of countries show a negative impact. However, the world economy is concentrated in a few, rich countries. The world average in Figure 6.3 counts dollars, rather than countries, let alone people.

Figure 6.4 suggests that poorer countries are more vulnerable to climate change than are richer countries. There are a few exceptions to this – such as Mongolia, which is poor but so cold that warming would bring benefits, and Singapore, which is rich but a low-lying island on the equator – but by and large the negative impacts of climate change are concentrated in the developing economies.

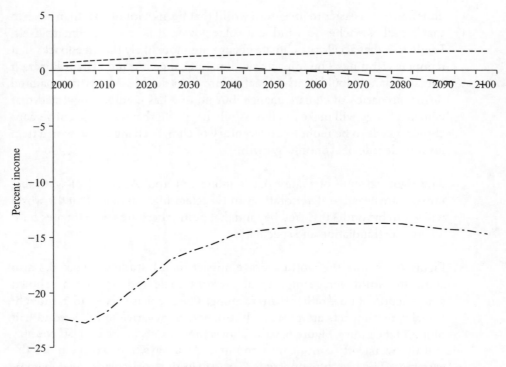

Figure 6.3 The economic impact of climate change at the world average and in the best-off and worst-off countries

There are three reasons for this. First, poorer countries are more exposed. Richer countries have a larger share of their economic activities in manufacturing and services, which are typically shielded (to a degree) against the vagaries of weather and hence climate change. Agriculture and water resources are far more important, relative to the size of the economy, in poorer countries.

Second, poorer countries tend to be in hotter places. This means that ecosystems are closer to their biophysical upper limits, and that there are no analogues for human behaviour and technology. Great Britain's future climate may become like Spain's current climate. The people of Britain would therefore adopt some of the habits of the people of Spain, and build their houses like the Spaniards do. If the hottest climate on the planet gets hotter still, there are no examples to copy from; new technologies will have to be invented, behaviour will have to be adjusted by trial and error.

Third, poorer countries tend to have a limited adaptive capacity. Adaptive capacity is the ability to adapt. It depends on a range of factors, such as the

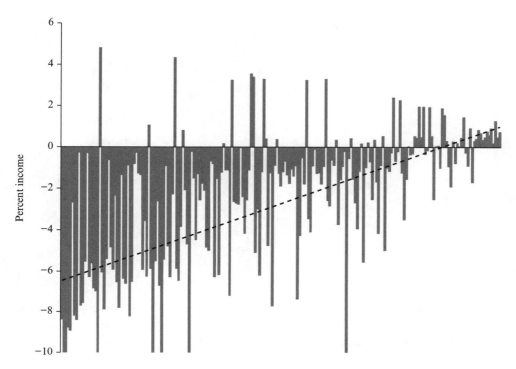

Figure 6.4 The economic impact of climate change in 2050 for all countries, which are ranked on per capita income in 1995

availability of technology and the ability to pay for those technologies. Sea level rise is a big problem if you do not know about dikes, or if you do but you cannot afford to build one. Adaptive capacity also depends on human and social capital. An ounce of prevention is worth a pound of cure, but prevention requires that you are able to recognize problems before they manifest themselves (i.e., predict the future) and that you are able to act on that knowledge (i.e., analytical capacity is connected to policy implementation). Furthermore, the powers that be need to care about the potential victims. A country's elite may be aware of the dangers of climate change and have the wherewithal to prevent the worst impacts, but if those impacts would fall on the politically and economically marginalized, the elite may chose to ignore the impacts.

6.3 Impacts and development**

The impacts of climate change are to a large degree determined by adaptation to climate change. Adaptation is constrained by adaptive capacity. The

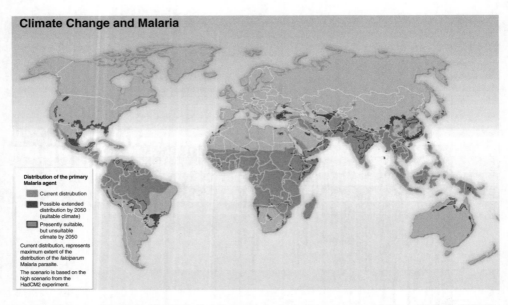

Climate Change and Malaria

Source: David J. Rogers and Sarah E. Randolph (2000), 'The global spread of malaria in a future, warmer world', *Science*, *289* (5485), 17631766. See http://www.grida.no/graphicslib/detail/climate-change-and-malaria-scenario-for-2050_bffe.

Figure 6.5 The impact of climate change on the malaria potential

components of adaptive capacity largely coincide with aspects of development. Therefore, future vulnerability to climate change will be very different from current vulnerability.

This is perhaps best illustrated with malaria. Figure 6.5 shows a map of projected changes in malaria due to climate change. A few areas see a decline. Most areas see an increase in the incidence of malaria, and the darker the colour the greater the increase. Malaria is introduced in the darkest areas. The mechanisms are as follows. Mosquitoes carry the disease. Mosquitoes are more active during warm weather. Mosquitoes need warm, still standing water to breed. A warmer and wetter world would thus see more mosquitoes. The malaria parasite develops faster in warm conditions. A warmer world would thus see more malaria.

Figure 6.6 shows the current and past distribution of malaria. In the lighter areas, all the natural conditions for malaria are met, but it does not occur. Malaria is now only present in the darkest areas. Affluence is the difference between the light and dark colours. Indeed, people in countries with an average income above $3000/person/year do not die from malaria (contracted in their home country). The mechanisms are as follows. Mosquitoes

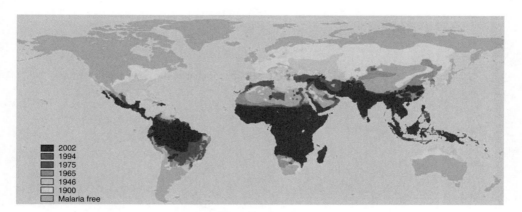

Source: Simon I. Hay, Carlos A. Guera, Andrew J. Tatem, Abdisalan M. Noor and Robert W. Snow (2004), 'The Global Distribution and Population at Risk of Malaria: Past, Present and Future', *Lancet Infectious Diseases*, 4 (6), 327–336.

Figure 6.6 The current and past distribution of malaria

need warm, still standing water to breed. First, draining of wetland, surfacing of roads and yards, and roofing of buildings have led to a dramatic decline in the number of small puddles of water in the developed world. Second, the large scale application of DDT in the 1950s killed many mosquitoes. Third, the life cycle of the malaria parasite requires both human and mosquito hosts. If a human is treated for malaria, she does not get sick. But she does not become infectious either. Herd immunity results if a sufficient number of people in a population take malaria medicine. A course in malaria medicine costs a few hundred dollars, a small fortune in poor countries and small change in rich countries. Malaria is thus a disease of both poverty and climate.

Figure 6.7 shows alternative projections of the future number of climate-change-induced malaria deaths. In blue, everything is kept constant except for the climate. The number of climate-change-induced malaria deaths increases from some 75,000 per year now to about 250,000 per year at the end of the century. Unfortunately, malaria records are not of sufficient quality to validate the model prediction of 75,000 climate deaths per year at present. In fact, the total estimate of malaria deaths was recently increased from 500,000 people per year to 1,000,000 people per year. The dotted lines indicate the uncertainty about the malaria model only (i.e., ignore the uncertainty about future greenhouse gas emissions or climate change). In the highest scenario, population growth is added and malaria numbers duly increase to about 750,000 deaths per year by 2100. This scenario is much quoted by environmentalists. In the third scenario, per capita income also grows. Malaria numbers first rise

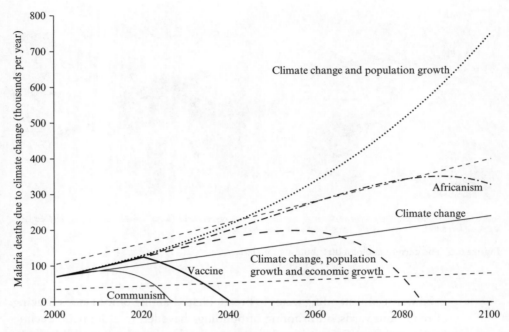

Figure 6.7 The impact of climate change on malaria for alternative scenarios: Climate change only (with, in the dashed lines, the 67% confidence interval including only the uncertainty about the malaria model); climate change and population growth; climate change, population growth and economic growth; the latter with public health spending typical for African countries; the same with public health spending typical for Communist countries; and the same with invention of a malaria vaccine

but later fall and malaria is eradicated around 2085 (in this model, with this parameterization, under this scenario). This scenario assumes that the global pattern of the relationship between health care and development holds for malaria. In two alternative scenarios, the communist pattern and the African pattern are used. Qualitatively, the results are the same: Malaria first goes up with warming and population growth before it falls with economic growth; in the communist pattern, the decline is sooner and faster, in the African pattern, the decline is later and slower. This illustrates the large uncertainty. As malaria is concentrated in Africa, the third scenario may be too optimistic. The final scenario adds another complication. It assumes that a malaria vaccine will be developed by 2020 (the deadline set by the Bill and Melinda Gates Foundation; a first vaccine was announced at the end of 2011). The vaccination campaigns against smallpox and polio took about 20 years. Using that number, malaria would be eradicated by 2040 – not just climate-change-induced malaria, but all malaria.

Figure 6.7 illustrates the relationships between the impacts of climate change and development, as well as the uncertainty about those relationships. It also begs a policy question. Malaria is one of the reasons for concern about climate change. Greenhouse gas emission reduction is not necessarily the best way to reduce the impact of climate change. Instead, money could be invested in medical technology or in public health care. We will return to these questions in Chapters 9 and 12.

6.4 Marginal economic impacts**

The focus above is on the total impact of climate change. From a policy perspective, the marginal impact is more relevant. This is because optimal climate policy (see Chapter 8) requires equating the marginal costs of greenhouse gas emission reduction (see Chapter 3) to its marginal benefits. Intuitively, climate change is a long-term, global problem. A single policy maker can only hope to change climate change by a little bit. The benefits of that are measured at the margin.

The marginal impact of greenhouse gas emissions is the damage done by emitting an additional tonne of, say, carbon dioxide. It is known as the marginal damage cost, and as the social cost of carbon (although I prefer to reserve that term for marginal damage cost estimates for a no climate policy scenario). It is the change in the net present value of the monetized impacts due to a small change in emissions, normalized by those emissions. Because of symmetry, the marginal damage of a small increase in emissions equals the marginal benefits of a small reduction in emissions. If the emissions trajectory is optimal, then the social cost of carbon equals the Pigou tax: The price we should put on greenhouse gas emissions if we wish to optimize net present welfare. Estimates of the social cost of carbon thus tell us what to do, how intensive climate policy should be, how much energy rises should be raised. It is a normative concept.

There have been many estimates of the marginal damage cost of carbon dioxide, the latest count standing at 759. At first sight, this is strange. Figure 6.2 shows only 17 estimates of the total impact of climate change. With 17 estimates of the total, how can there be 759 estimates of its first partial derivative? The answer is that there are 17 *comparative-static* estimates of the total impact of climate change *on the current economy* and *for a particular scenario*. The static results in Figure 6.2 need to be turned into dynamic ones by assuming a particular scenario for emissions and climate change, by assuming a scenario for development and the evolution of adaptive capacity, and by assuming functional forms of the relationship between impacts on the

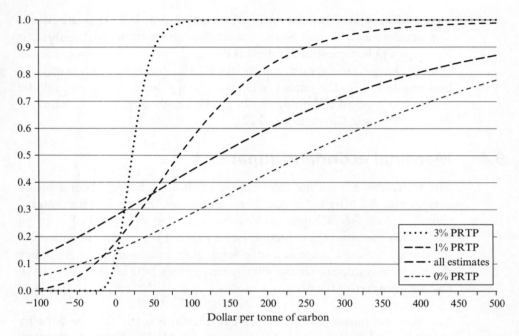

Figure 6.8 The cumulative distribution function of the social cost of carbon for all published studies and for all published studies that use a particular pure rate of time preference

one hand and climate and development on the other. Furthermore, impacts need to be aggregated over time, over space, and over states of the world (see Chapter 9). This introduces many additional degrees of freedom, which explains the proliferation of estimates of the marginal damage cost of carbon dioxide.

Figure 6.8 summarizes the many estimates in a cumulative density function (CDF). The CDF shows that, if all published estimates are considered, there is 42% chance that the marginal damage cost is less that $200/tC and a 58% chance that it is greater.

Figure 6.8 also illustrates the power of one of the most important parameters: The pure rate of time preference (PRTP). The PRTP is the utility discount rate. It measures how much we care about the future for the sake of it being then not now (see Chapter 9). The sample is split into four: Estimates that use a PRTP of 0%, 1% or 3% are shown, while other estimates (a handful only) are ignored. The lower the discount rate you use, the more you care about the future, the more you care about climate change, and the higher the marginal damage cost. The median estimate, for instance,

Table 6.1 The marginal damage costs of carbon dioxide emissions

	all	3%	1%	0%	SCC	Pigou
Mode	91	27	87	240	31	24
median	310	35	156	471	43	30
Mean	422	40	208	590	50	34
st dev	688	36	285	685	38	32
skewness	2	1	2	2	1	1
kurtosis	13	4	8	11	4	5

is \$35/tC for a 3% pure rate of time preference, \$156/tC for 1% rate, and \$471/tC for 0% rate.

Table 6.1 shows some of the characteristics. With a PRTP of 3%, a carbon price of \$40/tC can be justified. With a rate of 1%, a carbon price of \$208/tC passes the benefit–cost test.

The discount rate has another effect. The CDF for all estimates does not really converge to one, that is, there is a chance that the carbon tax should be greater than \$1,000/tC. This is entirely driven by the estimates that use a PRTP of 0%. For a higher rate, the CDF rapidly converges to one. This means that the discount rate not only discounts the impacts of climate change, but also the uncertainty about the impacts. This is intuitive. As we look further into the future, the uncertainty becomes ever larger. The discount rate curtails how far we look into the future, and thus how much uncertainty we have to contend with.

Figure 6.9 shows the same information as Figure 6.8 (for a pure rate of time preference of 3%), but now as a probability distribution function, the first partial derivative of the cumulative density function. Figure 6.9 splits the 3% PRTP sample into those studies that estimate the marginal damage cost along a no climate policy scenario, and those studies that impose a carbon tax equal to the marginal damage cost estimate. If the carbon tax equals the marginal damage cost, it is known as the Pigou tax (see Section 4.2). If a carbon tax is imposed, emissions fall and climate change is less of a problem. Therefore, the Pigou tax is less than the social cost of carbon: \$34/tC versus \$50/tC.

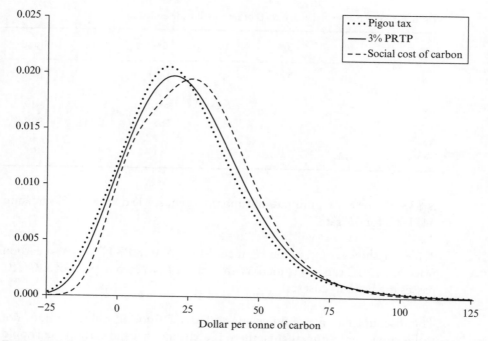

Figure 6.9 The probability density function of the social cost of carbon for all published studies that use a 3% pure rate of time preferences, for all studies that estimate the social costs of carbon and for all studies that estimate the Pigou tax

6.5 The growth rate of the marginal impact***

There are a number of studies of the evolution over time of the marginal damage costs of greenhouse gas emissions. Probability density functions of the results are displayed in Figure 6.10. If we take all studies, the mean growth rate of the marginal damage cost is 2.3% per year, with a standard deviation of 1.5%. If we take all studies that use a no-policy scenario, the mean growth rate of the social cost of carbon is 2.5% with a standard deviation of 1.8%. If we take all studies that use an optimal scenario, the mean growth rate of the Pigou tax is 2.1% with a standard deviation of 1.0%.

The difference in growth between the social cost of carbon and the Pigou tax is because climate policy affects climate change in the long run, but not in the short run. The Pigou tax is therefore not only lower than the social cost of carbon (cf. Figure 6.9), it also rises more slowly.

There is a sharp contrast between dynamic efficiency and dynamic cost-efficacy. In the latter case, the price of carbon should rise at a rate that is

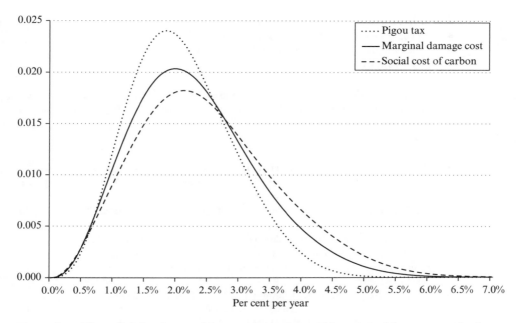

Figure 6.10 The probability density of the annual growth rate of the marginal damage cost of carbon dioxide emissions, the Pigou tax, and the social cost of carbon

about 0.6%[2] higher than the rate of discount (see Section 4.6). In the former case, the price of carbon should rise at some 2% per year.

 FURTHER READING

Every six years, Working Group II of the Intergovernmental Panel on Climate Change publishes a major assessment of the impacts of climate change. The information is layered, with a Summary for Policy Makers with high-level information, Technical Summaries with more detail, and multiple chapters with a lot of detail and references to the underlying literature. These reports can be found at: http://www.ipcc.ch/.

Samuel Fankhauser's *Valuing Climate Change: The Economics of the Greenhouse* (1995) remains the best introduction to the economic impact of climate change.

IDEAS/RePEc has a curated bibliography on this topic at http://biblio.repec.org/entry/tdd.html.

 EXERCISES

6.1 The statistical method to estimate the impacts of climate change uses the so-called ergodic assumption: It assumes that the relationship between welfare and climate that was estimated *over space* also holds *over time*. Formulate three objections to the ergodic assumption.

6.2 How will the marginal damage cost of carbon dioxide respond to:

- An increase in greenhouse gas emissions?
- An improvement in health care for infectious diseases?
- An improvement in health care for cardiovascular diseases?
- An increase in economic growth?

6.3 Suppose you have a budget of $100 million. You want to use this money to reduce the impacts of climate change on poor countries. How would you allocate the money over mitigation and adaptation?

6.4 Read and discuss:

- **G.W. Yohe and R.S.J. Tol (2002), 'Indicators for social and economic coping capacity – moving toward a working definition of adaptive capacity', *Global Environmental Change*, **12**, 25–40.
- **T.C. Schelling (2000), 'Intergenerational and international discounting', *Risk Analysis*, **20** (6), 833–837.
- ***W.N. Adger (2006), 'Vulnerability', *Global Environmental Change*, **16**, 268–281.
- ***R.S.J. Tol (2005), 'Emission abatement versus development as strategies to reduce vulnerability to climate change: An application of FUND', *Environment and Development Economics*, **10** (5), 615–629.
- ****J.P. Berrens et al. (2006), 'Information and effort in contingent valuation surveys: Application to global climate change using national Internet samples', *Journal of Environmental Economics and Management*, **47**, 331–363.
- ****W.K. Viscusi and R.J. Zeckhauser (2006), 'The perception and valuation of the risks of climate change: A rational and behavioral blend', *Climatic Change*, **77**, 151–177.

NOTES

1 Some island nations fare much worse.
2 The average atmospheric degradation over 100 years.

7
Climate and development

TWEET BOOK

- Climate change affects economic growth through its impact on productivity, labour force, and depreciation. #climateeconomics
- Negative impacts would decelerate growth. These indirect impacts are of the same size as the direct impacts. #climateeconomics
- The rational response to these impacts is to consume more, invest less and further decelerate growth. #climateeconomics
- Poor countries grow more slowly in hot years. Slow growth in Africa may be caused by secular decline in precipitation. #climateeconomics
- Some say climate and geography are economic destiny, others that climate and geography are not important. #climateeconomics
- Climate may contribute to trapping people in poverty, e.g., through infant mortality or via volatility and insecurity. #climateeconomics

7.1   Growth models**

Chapter 6 discusses the static impacts of climate change: Given a particular level of economic development, what would the effect of climate change be? However, climate change would also affect economic growth and development.

Climate change affects welfare in four different ways, three of which have ramifications for growth. Climate change may affect utility directly. For instance, climate change may drive a species to extinction. If this species has an existence value, and an existence value only, utility will fall – but the economy is not affected.

Climate change may also affect the size of the labour force, through changes in mortality, or its productivity, through changes in morbidity. This would have an impact on total output, and thus on investment and future output. Climate change may also affect productivity. These effects can be direct. For instance, crops may grow less well if it is hotter and drier. Traffic may be disrupted by extreme weather. Manual labour is harder in hot and humid climates. There are indirect effects on productivity too. Climate change would increase the demand for air conditioning. This changes the composition of supply too, in this case shifting towards a relatively unproductive sector (power generation). As productivity changes, so do output and investment. Finally, climate change may affect capital deprecation. More frequent floods, for instance, would wash away bridges, roads, and buildings. This implies that there is less capital and thus less output and investment. It also implies that more investment goes towards replacing capital and less towards expanding the capital stock.

Therefore, if the static impacts of climate change are negative, so are the impacts on economic growth. Calibrated models show that the indirect effect of climate change on welfare – lower income due to slower economic growth – is of similar size as the direct effect of climate change over the course of the 21st century.

The above assumes that people and companies do not adjust their savings and investment in response to climate change. Lower output does not necessarily mean lower investment if the savings' rate goes up. However, if anything, the savings' rate goes down in response to reduced output. First, for a fixed savings' rate, reduced output means reduced consumption. A higher savings' rate means lower consumption still. Second, if output is permanently reduced by climate change then so are the returns to investment. The optimal savings' rate thus falls.

Calibrated models indeed show that, with an endogenous savings' rate, economic growth decelerates further – if only slightly.

Such models also show that if more of economic growth is endogenous (to the model), the impact of climate change on growth is larger. In the canonical Solow model, 25% of economic growth is explained by capital accumulation (which is in the model) and 75% by technological progress (which is not in the model). If technological progress is included in the model – making it a so-called new growth model – then investments in R&D or human capital would fall just like investments in physical capital would. As a result, economic growth decelerates even further.

There is empirical evidence that the decline in rainfall in the 20th century partly explains that the economies of Sub-Saharan Africa have grown more slowly than those of other developing regions. In the second half of the 20th century, anomalously hot weather slowed down economic growth in poor countries, in both the agricultural and the industrial sectors.

7.2   Natural disasters***

Natural disasters have a number of effects on the economy. When disaster strikes, economic activity is disrupted and input factors are destroyed. Some disasters, such as floods and storms, particularly affect physical capital. Other disasters, such as epidemics, primarily affect the labour force. Disruption of economic activity shows up in GDP. Destruction of capital, on the other hand, does not.

After the disaster, there is recovery. The deaths are buried, debris cleared away, and houses and roads rebuilt. These are economic activities, and thus contribute to GDP.

Natural disasters thus neatly illustrate Bastiat's broken window fallacy: Destruction of the capital stock is not measured by GDP. Repair of the capital stock is. A naïve look at GDP growth rates may thus lead one to conclude that natural disasters are good for short-term economic growth. This is not the case. Natural disasters stimulate economic activity. Natural disasters do not improve welfare. GDP is a measure of economic activity. It is not a welfare indicator.

Natural disasters have different impacts at different phases of the business cycle. During a recession, the loss of input factors is less problematic as there is overcapacity anyway. The recovery phase is a Keynesian economic stimulus. During a boom, capacity is tight and lost inputs cannot readily be replaced. The demand stimulus from recovery may drive up inflation rather than output.

Natural disasters also have different impacts on different economies. Recovery requires resources. In developed economies, recovery is paid for by insurance, from household and company reserves, by the government, or by new loans from commercial lenders. In developing economies, contributions from these sources are limited, and recovery depends on support from informal networks and charity. Recovery from natural disasters is therefore slower, and sometimes much slower in developing countries than in developed economies.

Recovery replaces destroyed capital goods with new ones. Although the initial response is to restore things "exactly as they were", in fact replacements are often superior: New machinery would be state-of-the-art, new buildings better designed, and so on. This does not accelerate economic growth in the long run: The capital stock would be replaced anyway. Natural disasters force the hand of economic agents with regard to the timing of replacement investment. Discretionary timing would be preferred.

The impact of natural disasters on the economy in the short term is therefore mixed, but probably negative on net. The same is true in the long term. If there is a risk of natural disasters, resources are diverted to protective measures, be they financial (e.g., insurance) or physical (e.g., dikes). If the risk were zero, those resources could be used for consumption or for investment in productive assets.

The empirical evidence on the impact of natural disasters is mixed, partly because of the measurement errors noted above – GDP is a poor indicator but most widely reported – and partly because of endogeneity issues – richer societies are better able to shield themselves from natural disasters. There is some evidence that natural disasters disproportionally affect the growth rate of poor countries.

7.3   Poverty traps**

Poverty is concentrated in the tropics and subtropics. Figure 7.1 shows the standard of living – per capita income measured in purchasing parity dollar – for a large number of countries at nine different points in time. Throughout history, very cold and very hot countries appear to be at a disadvantage. Over time, the point of gravity of the world economy shifted from the Middle East and Mediterranean to cooler places, but tropical countries have always been the poorest. Colder countries appear to have lost their disadvantage earlier than hotter countries (excepting those with plenty of oil and gas).

Although the correlation is clear, causation is disputed. Some argue that climate and geography are destiny. This was a popular theory in centuries past. Jared Diamond is the main current proponent of this hypothesis. Others, including Jeffrey Sachs, argue that climate and geography are contributing factors to (under)development, influential but not predominant. Yet others, including Daron Acemoglu, argue that climate and geography were important in the past in shaping institutions (education, rule of law, etc.) but that institutions explain the current pattern of development. Others still, including William Easterly, argue that institutions are the only thing that matters.

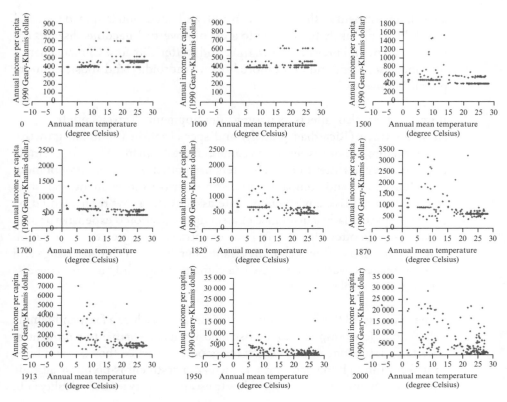

Figure 7.1 Standard of living as a function of temperature in the years 0, 1000, 1500, 1700, 1820, 1870, 1913, 1950 and 2000

The two extreme positions – climate and geography are the only things that matter, or do not matter at all – cannot seriously be maintained. Diamond's hypothesis is easily dismissed when comparing the two halves of Hispaniola and Korea. Acemoglu's position – that climate used to be a factor but not anymore – cannot be supported either. Climate indisputably affects agriculture, energy, health, and tourism. That also falsifies Easterly's hypothesis that the only things that matter to humans are other humans. The question is not whether but to what extent climate determines the pattern of development. Could it be that climate is a contributing factor to the probability that people are trapped in poverty?

Diseases such as malaria and diarrhoea impair children's cognitive and physical development. This leads to poverty in their later life so that there are limited means to protect their own children against these diseases. Furthermore, high infant mortality may induce parents to have many children, and risk-averse parents to have more children than they really want. As a result, investment in the children's health and education is spread thin and the children are likely to grow up to a life of poverty. Infectious diseases are more virulent and prevalent in warmer and wetter climates. Climate change may increase infant and child mortality and morbidity and thus trap more people in poverty.

Infrastructure also affects economic development. Travel and transport allow for trade (Ricardian growth) and specialization (Smithian growth). Infrastructure is more expensive in some climates than in others, for example because of repairs after floods. Disasters have other effects on development too. Households and companies trade-off returns to investment against its safety. In a high-risk environment, safe assets with a low return would be preferred, particularly if insurance or asset diversification is expensive or una-available – as is often the case in poor countries. Slow growth is the result. The jury is still out, though, on the relationship between climate change and weather-related disasters, and between climate change and human-made dis-asters (such as violent conflict).

Some have argued that highly volatile environments induce a feast-and-famine culture. A rational response to the risk of losing it all, whether to a drought or a warlord, could indeed be to enjoy the good times while they last and hope to make it through the bad times. If such an investment-is-pointless attitude becomes engrained, people may be slow to escape poverty even if volatility falls.

In sum, climate affects development, and may trap people in poverty. The upshot is that the impacts of climate change may be (much) larger than commonly believed. The welfare impacts of growing at a small rate versus not growing at all are large when accumulated over a century or more.

There is limited empirical evidence that hot and wet conditions and large variability in rainfall reduce long-term growth in poor countries (but not in hot ones) and increase the probability of being poor.

 FURTHER READING

IDEAS/RePEc has a curated bibliography on this topic at http://biblio.repec.org/entry/tde.html.

 EXERCISES

7.1 Construct a Solow growth model of a one-sector, closed economy, assuming a Cobb–Douglas production function in labour and capital. Assume that the capital stock is in its steady-state value (either by running the model for 100 years without changing anything, or by solving the steady-state analytically). Assume, in turn, that climate change

- reduces total factor productivity by 10%
- increases depreciation by 10%
- reduces labour supply by 10%

What are the implications for per capita consumption after 1 year, 10 years, 100 years?

7.2 Read and discuss:

- **W. Easterly and R. Levine (2003), 'Tropics, germs and crops: How endowments influence economic development', *Journal of Monetary Economics*, **50**, 3–39.
- **S. Barrios, L. Bertinelli and E. Strobl (2011), 'Trends in rainfall and economic growth in Africa: A neglected cause of the African growth tragedy', *Review of Economics and Statistics*, **92**, 350–366.
- ***S. Fankhauser and R.S.J. Tol (2005), 'On climate change and growth', *Resource and Energy Economics*, **27**, 1–17.
- ***D. Acemoglu, S. Johnson and J.A. Robinson (2001), 'The colonial origins of comparative development: An empirical investigation', *American Economic Review*, **91**, 1369–1401.
- ****M. Dell, B.F. Jones and B.A. Olken (2012), 'Temperature shocks and economic growth: Evidence from the last half century', *American Economic Journal: Macroeconomics*, **4** (3), 66–95.
- ****D.E. Bloom, D. Canning and J. Sevilla (2003), 'Geography and poverty traps', *Journal of Economic Growth*, **8**, 355–378.
- ****O. Galor and D.N. Weil (1996), 'The gender gap, fertility and growth', *American Economic Review*, **86**, 374–387.

8

Optimal climate policy

TWEET BOOK

- Article 2 of the UN Framework Convention on Climate Change calls for stabilization of the atmospheric concentration of CO_2. #climateeconomics
- Part of CO_2 emissions stay in the atmosphere forever – or rather, are removed at the rate at which rocks grow. #climateeconomics
- Thus stabilization of the atmospheric concentration of CO_2 implies that CO_2 emissions have to go to zero. #climateeconomics
- In the social optimum, the marginal net present costs of a policy equal its marginal net present benefits. #climateeconomics
- The marginal costs of complete emission reduction are high. Eliminating all emissions is very difficult. #climateeconomics
- The incremental benefits of complete emission reduction are small. A little bit of climate change does little damage. #climateeconomics
- It cannot be optimal to reduce greenhouse gas emissions to zero, so it cannot be optimal to stabilize CO_2 concentration. #climateeconomics
- Estimates of the abatement costs and benefits point to an optimal emission reduction effort that is modest. #climateeconomics
- Whereas all credible studies agree that emission reduction should start slow and stop short of 100% in the long-term. #climateeconomics
- The magnitude of the "modest" effort in the medium term is very sensitive to a large number of assumptions. #climateeconomics
- 100% emission reduction is justified if there is a perfect and permanent substitute for fossil fuels at a reasonable price. #climateeconomics

8.1   The ultimate target**

Article 2 of the United Nations Framework Convention on Climate Change (UNFCCC), ratified by almost all countries, states that

> the ultimate objective of [. . .] is to achieve [. . .] stabilization of greenhouse gas concentrations in the atmosphere at a level that would prevent dangerous anthropogenic interference with the climate system. Such a level should be

achieved within a time-frame sufficient to allow ecosystems to adapt naturally, to ensure that food production is not threatened and to enable economic development to proceed in a sustainable manner.

Anyone can read anything into the second sentence. It is waffle. Appealing but vapid language makes great diplomacy; everyone can sign up. The first sentence seems to be of a similar nature. How can anyone object to avoiding (otherwise unspecified) danger? The word "stabilization", however, is all but innocuous.

Let us consider a simple stock model:

$$S_t = (1 - \delta)S_{t-1} + E_{t-1} \qquad (8.1)$$

where S_t is the stock at time t, E is emissions, and δ is the rate of degradation. Equation (8.1) is a first-order difference equation. It describes a geometric process.

Stabilization requires that the stock does not change: $S_t = S_{t-1}$. Then

$$S = (1 - \delta)S + E \Leftrightarrow \delta S = E \Leftrightarrow S = \frac{E}{\delta} \qquad (8.2)$$

That is, if emissions are stabilized – $E_t = E_{t-1}$ – then concentrations stabilize too at a level that is inversely proportional to the rate of degradation of emissions in the atmosphere. In this case, stabilization of concentrations implies that emissions be stabilized *at any level*. Article 2 appears void.

Figure 1.6 shows a stylized representation of the carbon cycle. Carbon dioxide is removed from the atmosphere by a number of processes. One of these is the weathering of rock, which is a very slow process. Mathematically, it is best to think about carbon dioxide in the atmosphere as five separate stocks.

$$S_t = \sum_{i=1}^{5} S_{i,t} \qquad (8.3)$$

with

$$S_{i,t} = (1 - \delta_i)S_{i,t-1} + \alpha_i E_{t-1}; \sum_{i=1}^{5} \alpha_i = 1 \qquad (8.4)$$

Atmospheric stabilization then requires stabilization of each of the five sub-concentrations. However, atmospheric degradation in one of the five components is by a geological process (rock weathering) at a geological time scale. At a human time scale, there is no degradation at all: $\delta_i=0$ for $i=5$; $\alpha_i=0.13$. That is, about 13% of anthropogenic carbon dioxide emissions stay in the atmosphere forever.

Figure 8.1 illustrates this for past emissions of carbon dioxide. The darkest colour represents the background concentration. In slightly lighter grey are the "permanent" additions to the atmosphere. The lighter tones represent those parts of the carbon dioxide concentration that will eventually disappear. Figure 8.2 repeats this for scenarios of future emissions. The solid lines are the actual concentrations, the dashed lines the "permanent" parts.

The permanence of emissions poses a problem. In Equation (8.2), the stable concentration is inversely proportional to the rate of degradation. Dividing by zero is not possible. However, there is a solution for $\delta = 0$. If $E=0$, concentrations stabilize.

The first sentence of Article 2 of the UN Framework Convention on Climate Change is not vacuous. In fact, it is radical. Stabilization of the atmospheric concentration of carbon dioxide requires that emissions are reduced to zero. Almost all countries are under a legal obligation to reduce their emissions by 100%.

There is no evidence that suggests that the people who drafted Article 2 were aware of this. The people who ratified the UNFCCC probably did not realize the implications either. Indeed, politicians regularly refer to an 80% emission reduction goal in the long run, with the greener ones opting for 90%. International law says it is 100%.

8.2 Benefit–cost analysis*

Benefit–cost analysis seeks to find the best course of action. If there are a finite number of options, then you should estimate the costs and benefits of each option. You should discard the options that have greater costs than benefits. The remaining options should be ranked on the basis of the benefit–cost ratio. You should fund the projects with the highest benefit–cost ratio until you run out of budget. You may consider borrowing money to fund (some of) the remaining projects.

Benefit–cost analysis works differently if there is a continuum of actions. A carbon tax, for example, can take any value (even if politicians tend to think in rounded numbers). In this case, benefit–cost analysis seeks to maximize

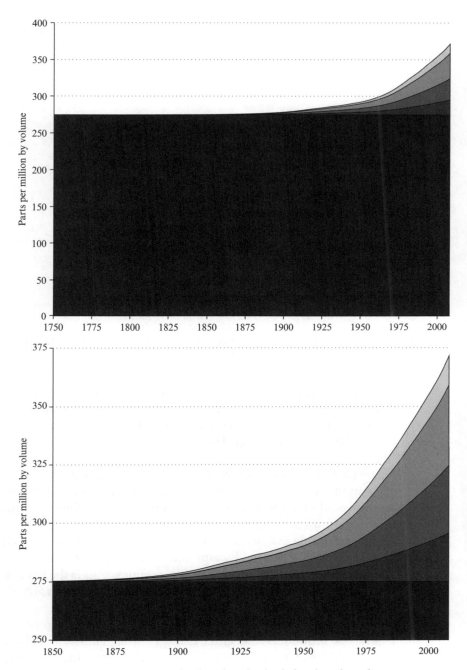

Figure 8.1 The atmospheric concentration of carbon dioxide; the darker the colour, the more permanent the concentration; the bottom panel enlarges aspects of the top panel

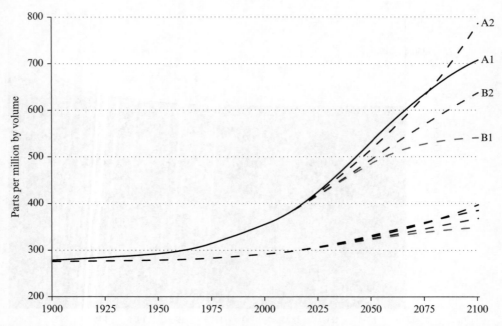

Figure 8.2 The atmospheric concentration of carbon dioxide according to four SRES scenarios; the dotted lines give the permanent parts of the concentrations

the objective function – net benefits, or benefits minus costs. The maximum is found by differentiation. The first order condition is that the first partial derivative of net benefits to the control variable be equal to zero. Rearranging, marginal benefits should equal marginal costs.

Figures 8.3, 8.4 and 8.5 derive this graphically. Figure 8.3 shows the gross gains from emissions. These increase first with emissions, but fall later. If not, it would be optimal to emit an infinite amount. This is intuitive: We heat our homes to a comfortable level, but not beyond because that would be uncomfortable and cost money. We travel to where we need to be, but not beyond because that would waste time and money. The private optimum emissions are where the curve is at its maximum. This is the point at which the slope of the curve – the marginal gains – is zero. This is also shown in Figure 8.3.

Figure 8.4 introduces the gross damages from emissions. The more is emitted, the greater the damage. Figure 8.4 also shows the net gains, that is, the gross gains minus the gross damages. Like the gross gains, the net gains first increase and then decrease with emissions. There is a maximum for the net gains, but that lies to the left of the maximum for the gross gains.

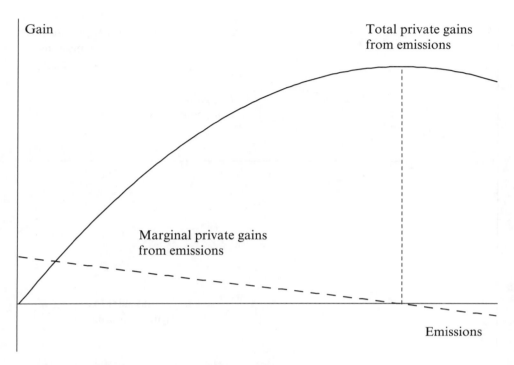

Figure 8.3 Optimal emissions if there are no external costs

Figure 8.5 shows the slopes of the curves of Figure 8.4, or the marginal gains and losses. Note that the marginal damages have been reflected in the x-axis. This is the graphical equivalent of the algebraic move to the other side of the equation. The net gain is maximum where the marginal net gain is zero. The marginal net gain is zero where the marginal cost equals the marginal benefit.

In terms of calculus, consider

$$\max_{E} G = D(E) - B(E) \tag{8.5}$$

where W denotes net gains, E are emissions, D are the damages of emissions and B are the benefits of emissions. The emissions are the control variable. That is, emissions are chosen so as to maximize net gains. Then the first-order conditions are

$$\frac{\partial G}{\partial E} = 0 \Leftrightarrow \frac{\partial D}{\partial E} - \frac{\partial B}{\partial E} = 0 \Leftrightarrow \frac{\partial D}{\partial E} = \frac{\partial B}{\partial E} \tag{8.6}$$

That is, the marginal benefits of emissions should equal the marginal damages of emissions.

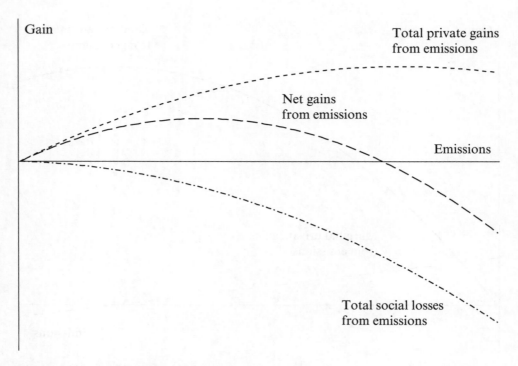

Figure 8.4 Costs and benefits of emissions

Climate change is a dynamic problem, however. Therefore, we need to rewrite Equation (8.5)

$$\max_{E_0, E_1, \ldots} \sum_t \frac{G_t}{(1 + r)^t} = \sum_t \frac{D_t(E_t, E_{t-1}, \ldots, E_0) - B_t(E_t)}{(1 + r)^t} \quad (8.7)$$

where r is the discount rate. That is, we maximize the net present value of the gains rather than the net gains. For simplicity, we assume that the benefits of emissions are instantaneous: The benefits at time t only depend on the emissions at time t. By contrast, the damages of emissions depend on emissions in all previous periods.

The maximization problem is structurally different: Instead of choosing the level of emissions as in Equation (8.6), the level of emissions needs to be chosen simultaneously at every point in time. There are therefore many first order conditions:

$$\sum_s \frac{1}{(1 + r)^s} \frac{\partial D_{t+s}}{\partial E_t} = \frac{\partial B_t}{\partial E_t} \, \forall t \quad (8.8)$$

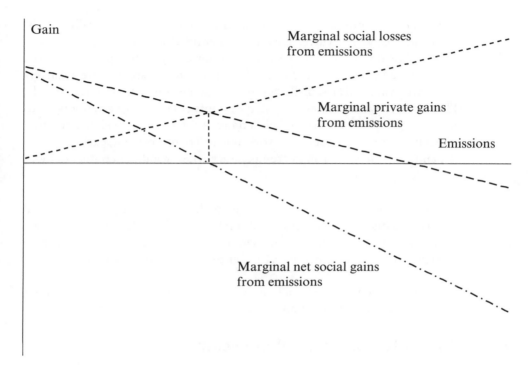

Gain

Marginal social losses
from emissions

Marginal private gains
from emissions

Emissions

Marginal net social gains
from emissions

Figure 8.5 Optimal emissions with external costs

That is, at every point in time, the marginal benefits of emissions should equal the net present value of the marginal damages of emissions.

That said, Figure 8.5 illustrates Equation (8.8) as well as Equation (8.6): Costs and (net present) benefits should be equal at the margin (simultaneously at every point in time).

Figure 8.5 also illustrates two fundamental insights from benefit–cost analysis. First, if there are damages from emissions, then it is optimal to reduce emissions. In fact, it is relatively cheap to reduce emissions by the first bit while the benefits of the first bit of emission reduction are relatively high. Therefore, benefit–cost analysis calls for action that goes beyond token emission reduction. Second, it is relatively expensive to reduce the final bit of emissions while the benefits of reducing the final bit are relatively low. Therefore, benefit–cost analysis rarely calls for a complete elimination of emissions.

Applied to climate change, greenhouse gas emissions can be reduced by a little bit without much of a bother. Energy use is often wasteful or the result of perverse incentives. Eliminating all emissions is disruptive. While technical but costly alternatives have been identified for most applications of carbon dioxide, this is not (yet) the case for every niche application (e.g., in space travel and the military). Alternative energy sources are relatively cheap when applied at a small scale, but much more expensive at a large scale. Photovoltaic panels can be mounted on roofs (that is, with a zero opportunity cost for space) but roof space is finite. A little wind power can be easily integrated into an electricity network, but grid reinforcement, back up capacity, and frequency regulators are necessary when wind penetration is more than a few percent.

Vice versa, uncontrolled climate change can do a lot of damage and the initial emission reductions thus bring substantial benefits. However, as climate change is reduced further and further, those benefits fall. The benefits from reducing global warming from 0.1°C per century to 0.01°C are small.

Therefore, a benefit–cost analysis will not recommend a 100% emission reduction. Yet, a 100% emission reduction is required by international law.

8.3   Estimates of optimal emission reduction**

Estimates of the marginal costs of greenhouse gas emission reduction were shown in Chapter 3. Estimates of the marginal benefits were shown in Chapter 6. A cursory comparison reveals that while emission reduction can be justified, deep cuts cannot.

This is confirmed by Figure 8.6 and Figure 8.7. It shows results of the DICE model, developed by William Nordhaus of Yale University. Figure 8.6 shows the carbon tax that maximizes global welfare, and the corresponding emission control rate. Figure 8.7 shows the corresponding atmospheric concentrations of carbon dioxide.

Essentially, the DICE model answers the question "if the world were ruled by a benevolent dictator, a philosopher-queen, what would she do?" The answer is a little bit of emission reduction at first, more later, but not enough to stabilize the carbon dioxide concentration.

This result depends, of course, on the many assumptions made in the DICE model. These assumptions are uncertain, and some are controversial. Nordhaus first published the bottom-line conclusion in 1991. Many researchers have tried to overturn the result, which came as a shock then and still does annoy people.

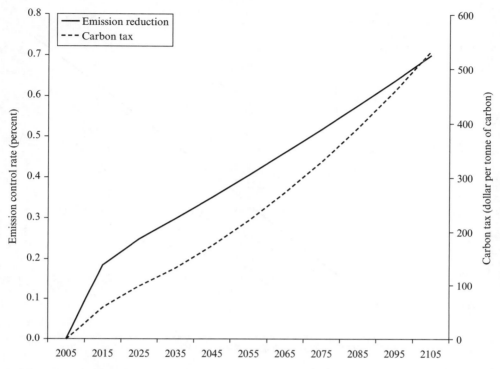

Figure 8.6 Optimal emission control and carbon tax

The following results have emerged since 1991. The first part of the conclusion is robust. In the short run, emission reduction should be modest regardless of the assumptions that you make or the way in which you structure the model or the problem. Let us assume that a carbon tax is the instrument of choice for climate policy. A carbon tax is there to change behaviour. However, a substantial part of our behaviour with regard to energy use is fixed. For households, energy use depends on the houses we will live in, the cars we drive, and where we go to school or work. A carbon tax may induce us to buy a different car or move closer to work – but only when we had planned to replace our car or move anyway. In the meantime, a carbon tax imposes a cost without any gain. Similarly, corporate energy use is determined to a large degree by machinery, buildings, and locations – things that change slowly. Rapid emission reduction would require that we discard perfectly good durable and production goods. Capital destruction would entail a large cost.

The second part of the conclusion is very sensitive to assumptions. The rate of acceleration of climate policy can take a wide range of values, depending on the model parameter and the structure of the model.

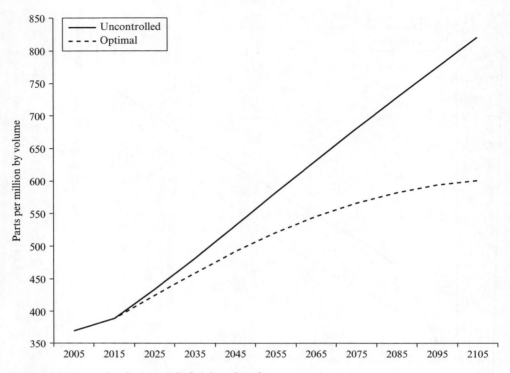

Figure 8.7 Optimal and uncontrolled carbon dioxide concentration

However, the third part of the conclusion is again robust. As argued in Section 8.3, it is hard to imagine that a benefit–cost analysis would ever lead to a 100% emission reduction, and a 100% emission reduction is needed to stabilize the atmospheric concentration of carbon dioxide.

There is one exception to this. If there is a backstop technology – a carbon-neutral energy source that is so abundant that its supply curve is flat – and if the economy is hysteretic, that is, if the economy would not leave its carbon-neutral state without active policy intervention – then benefit–cost analysis would justify complete emission reduction. If the social cost of carbon would be high enough to trigger the deployment of the backstop technology, then the economy would decarbonize and never look back. These are big ifs, though.

8.4   Trade-offs between greenhouse gases****

The first-order condition for optimal emission reduction is that the marginal costs of greenhouse gas emission reduction equals its marginal benefits. If carbon dioxide were the only greenhouse gas, the first-order condition establishes a price for its emission. There are many greenhouse gases, in fact, each with its own first-order conditions and its own price.

Instead of looking at the absolute prices of all greenhouse gases, it is more instructive to consider the absolute price of carbon dioxide (see Sections 3.2 and 6.5) and the relative prices of all other greenhouse gases.

Let us denote the marginal damage of carbon dioxide emissions by MD_0

$$MD_0 = \sum_t \frac{1}{(1+r)^t} \frac{\partial D_t}{\partial E_0} = \sum_t \frac{1}{(1+r)^t} \frac{\partial D_t}{\partial T_t} \frac{\partial T_t}{\partial E_0}$$

$$= \sum_t \frac{1}{(1+r)^t} \frac{\partial D_t}{\partial T_t} \sum_s \frac{\partial T_t}{\partial F_{t-s}} \frac{\partial F_{t-s}}{\partial C_{0,t-s}} \frac{\partial C_{0,t-s}}{\partial E_0} \tag{8.9}$$

The first term on the right-hand side is as above – compare with Equation (8.8). The second term recognizes that the impact of climate change does not directly depend on emissions E, but rather via climate change T. The third and final term has that climate change T at time t depends on radiative forcing F in previous periods, which in turn depends on concentrations C and emissions E.

The relative price of greenhouse gas i is then MD_0

$$\frac{MD_i}{MD_0} = \frac{\sum_t \frac{1}{(1+r)^t} \frac{\partial D_t}{\partial T_t} \sum_s \frac{\partial T_t}{\partial F_{t-s}} \frac{\partial F_{t-s}}{\partial C_{i,t-s}} \frac{\partial C_{i,t-s}}{\partial E_i}}{\sum_t \frac{1}{(1+r)^t} \frac{\partial D_t}{\partial T_t} \sum_s \frac{\partial T_t}{\partial F_{t-s}} \frac{\partial F_{t-s}}{\partial C_{0,t-s}} \frac{\partial C_{0,t-s}}{\partial E_0}} \tag{8.10}$$

The nominator and the denominator of Equation (8.10) have several elements in common – particularly, the discount rate and the damage function – suggesting that the uncertainty about the relative price of greenhouse gas i is smaller than the uncertainty about its absolute price.

There are a large number of assumptions in Equation (8.10). One can set the discount rate to zero $r=0$. One can assume that impacts of climate change are proportional to temperature: $\partial D_t/\partial T_t = \alpha$. Then

$$\frac{MD_i}{MD_0} = \frac{\sum_t \sum_s \dfrac{\partial T_t}{\partial F_{t-s}} \dfrac{\partial F_{t-s}}{\partial C_{i,t-s}} \dfrac{\partial C_{i,t-s}}{\partial E_i}}{\sum_t \sum_s \dfrac{\partial T_t}{\partial F_{t-s}} \dfrac{\partial F_{t-s}}{\partial C_{0,t-s}} \dfrac{\partial C_{0,t-s}}{\partial E_0}} \tag{8.11}$$

One can assume that climate change is proportional to radiative forcing: $\partial T_t/\partial F_s = \beta$. One can further assume that concentrations are unchanged. Finally, one can ignore impacts after $t = H$. Then

$$\frac{MD_i}{MD_0} = \frac{\sum_{t=0}^{H} \dfrac{\partial F_t}{\partial C_i} \dfrac{\partial C_i}{\partial E_i}}{\sum_{t=0}^{H} \dfrac{\partial F_t}{\partial C_0} \dfrac{\partial C_0}{\partial E_0}} \tag{8.12}$$

The assumptions leading from (8.10) to (8.12) seem restrictive: No discounting; linear impacts; linear climate change; constant concentrations; finite time horizon. Equation (8.12) is known as the Global Warming Potential. The Kyoto Protocol of the United Nations Framework Convention for Climate Change specifies that the relative prices of greenhouse gas be based on Equation (8.12) (using parameter values that were conventional wisdom in 2001). The Kyoto Protocol thus introduces a distortion in the relative prices of greenhouse gas.

The Kyoto Protocol also introduces an internal inconsistency. Equation (8.10) gives the relative price in the optimum. Equation (8.12) is a special case of that equation, and therefore still based on benefit–cost reasoning. The UNFCCC and its Kyoto Protocol explicitly reject such reasoning. Instead, a target is set that should be met at the lowest possible cost. This is a cost-effectiveness analysis, rather than a cost–benefit analysis.

For a single gas, a cost-effective emission trajectory solves

$$\min_{E_{0,0}, E_{0,1}, \dots} \sum_t B_t (1 + r)^{-t} \text{ s.t. } T_H \leq \overline{T} \tag{8.13}$$

The first-order conditions are

$$\frac{\partial B_t}{\partial E_{0,t}} (1 + r)^{-t} = \lambda \frac{\partial T_H}{\partial E_{0,t}} \quad \forall t \tag{8.14}$$

That is, the marginal present emission reduction cost (on the left hand side) equal the shadow price of the intertemporal constraint λ times the marginal contribution to meeting that constraint.

For multiple gases, the first-order conditions are similar. There is an extra index to denote the various gases. The constraint is shared over time and between gases.

The relative price of greenhouse gas emissions then becomes

$$\frac{\dfrac{\partial B_t}{\partial E_{i,t}}}{\dfrac{\partial B_t}{\partial E_{0,t}}} = \frac{\lambda \dfrac{\partial T_H}{\partial E_{i,t}}(1+r)^t}{\lambda \dfrac{\partial T_H}{\partial E_{0,t}}(1+r)^t} = \frac{\dfrac{\partial T_H}{\partial E_{i,t}}}{\dfrac{\partial T_H}{\partial E_{0,t}}} \quad \forall i,t \qquad (8.15)$$

That is, the cost-effective ratio of marginal emission reduction costs equals the ratio of marginal contributions to the constraint.

Equation (8.15) is very different from Equations (8.12) and (8.10). The latter two are integrals over time whereas the former is instantaneous at the time the constraint bites. The difference is most pronounced for short-lived gases. Methane, for instance, has an atmospheric half-life of a decade or so, compared with carbon dioxide which stays in the atmosphere for decades and centuries. Methane emission reduction therefore contributes to reduced climate change in the short run – which may count as a substantial benefit with a high discount rate – but not in the long run. A benefit–cost analysis thus favours methane emission reduction now, whereas a cost-effectiveness analysis does not.

 FURTHER READING

William Nordhaus' *Managing the Global Commons: Economics of Climate Change* (1994), Nordhaus and Joseph Boyer's *Warming and the World: Economic Models of Global Warming* (2000), and Nordhaus' *A Question of Balance: Weighing the Options on Global Warming Policies* (2008) set the standard for analyses of optimal climate policy.

IDEAS/RePEc has a curated bibliography on this topic at http://biblio.repec.org/entry/tdf.html.

 EXERCISES

8.1 The European Union argues that the carbon dioxide concentration should not exceed 450 ppm. Work out the global emissions budget. Work out the global emissions budget if the constraint holds only in the very long run. (Hint: Past emissions and a simple carbon cycle model are at https://sites.google.com/site/climateconomics/mliam under MLIAM01.)

8.2 In the DICE model, Nordhaus assumes that the impact of climate change is proportional to $0.28T^2$, where T is the global mean surface air temperature (in degrees Celsius, in deviation from pre-industrial times). What happens to optimal climate policy if the impact function is $-4.33T+1.92T^2$ instead (as suggested by Tol) or $-0.348T^2+0.0109T^6$ (as suggested by Weitzman)? (Hint: The model code is at https://sites.google.com/site/climateconomics/data/08-optimal-climate-policy under DICE2007.)

8.3 In the DICE model, Nordhaus assumes that there is a backstop technology that can provide carbon-free energy at a fixed cost. In the standard model, the cost is set at \$1260/tC in 2005 (and falling over time). What happens to optimal climate policy if this is set at \$630/tC and \$2520/tC? (Hint: The model code is at https://sites.google.com/site/climateconomics/data/08-optimal-climate-policy under DICE2007.)

8.4 Read and discuss:

- **T.C. Schelling (1995), 'Intergenerational discounting', *Energy Policy*, **23** (4/5), 395–401.
- **T.C. Schelling (2000), 'Intergenerational and international discounting', *Risk Analysis*, **20** (6), 833–837.
- **T.M.L. Wigley, R.G. Richels and J.A. Edmonds (1996), 'Economic and environmental choices in the stabilization of atmospheric CO_2', *Nature*, **379**, 240–243.
- ***R.S.J. Tol (2005), 'Emission abatement versus development as strategies to reduce vulnerability to climate change: an application of FUND', *Environment and Development Economics*, **10**, 615–629.
- ***L.H. Goulder and K. Mathai (2000), 'Optimal CO_2 abatement in the presence of induced technological change', *Journal of Environmental Economics and Management*, **39**, 1–18.
- ****D. Burtraw et al. (2003), 'Ancillary benefits of reduced air pollution in the US from moderate greenhouse gas mitigation policies in the electricity sector', *Journal of Environmental Economics and Management*, **45**, 650–673.
- ****N.H. Stern (2008), 'The economics of climate change', *American Economic Review*, **98**, 1–37.

9

Discounting, equity, uncertainty

TWEET BOOK

- You'd only care about climate change if you care about the distant future, far-away lands, remote probabilities. #climateeconomics
- We discount the future because we are impatient and because we expect to become richer and happier. #climateeconomics
- The Ramsey Rule has these three components: pure time preference and consumption growth transformed to utility growth. #climateeconomics
- Ramsey Rule describes discount rate of consumers, in equilibrium equal to the interest rate paid on the capital market. #climateeconomics
- The total and marginal net present impact of climate change rise sharply with a falling discount rate. #climateeconomics
- It is wrong to add up monetized impacts because a dollar to a poor woman is not a dollar to a rich woman. #climateeconomics
- Equity weights correct for this, putting more weight on the impacts in poorer countries but keeping their values as is. #climateeconomics
- Equity weights tend to increase the global impact, as poorer countries tend to be more vulnerable to climate change. #climateeconomics
- It is wrong to use the average monetized impact because the more severe scenarios have a larger effect on utility. #climateeconomics
- The risk premium corrects for this, putting more weight on low probability, high impact scenarios. #climateeconomics
- The risk premium tends to be positive because climate change risks are skewed towards bad outcomes. #climateeconomics
- Welfare functions can be based on ethical reasons or on empirical evidence, but should be consistent across policies. #climateeconomics
- Ethical welfare functions are undemocratic. Empirical evidence is contradictory and difficult to interpret. #climateeconomics

9.1 Time discounting: Derivation of the Ramsey rule***

Climate change is a long-term, global, uncertain problem. If you do not care about the distant future, far-away lands, and remote probabilities, then climate change is of little concern.

The consumption rate of discount r measures trade-offs over time. The consumption rate of discount is defined as that rate that leaves you indifferent between a reduction in consumption now in return for an increase in consumption later; or between an increase in consumption now in return for a later decrease.

The consumption rate of discount is thus a rate of change: $(X(t)-X(0))/X(0)$ between now ($t=0$) and some later time t. It is a rate of change, not of consumption C, but of a change in consumption, dC. Thus the consumption rate of discount is defined as

$$e^{rt}:=\frac{dC_t}{dC_0} \Leftrightarrow e^{rt}dC_0 = dC_t \Leftrightarrow dC_0 = e^{-rt}dC_t \qquad (9.1)$$

If we set $dC_t=1$ then it is obvious that e^{-rt} is the discount factor: the present value of money received in the future.

Equation (9.1) is not complete because it omits the indifference requirement. That is, the net present welfare W is unaffected by the shift in consumption from now to then. Or (approximately) $W(C_0,C_1,\ldots C_{t-1},C_t,C_{t+1},\ldots)= W(C_0+dC_0,C_1,\ldots C_{t-1},C_t-dC_t,C_{t+1},\ldots)$. Actually, this condition should hold at the margin. That is, the total derivative should be equal to zero.

If we assume that net present welfare function is as follows

$$W = \int_t U(C_t)e^{-\rho t}dt \qquad (9.2)$$

where U is instantaneous utility and ρ the utility rate of discount, then the condition is

$$dW = \frac{\partial U}{\partial C_0}dC_0 - e^{-\rho t}\frac{\partial U}{\partial C_t}dC_t = 0 \Leftrightarrow \frac{\partial U}{\partial C_0}dC_0 = e^{-\rho t}\frac{\partial U}{\partial C_t}dC_t$$

$$\Leftrightarrow \frac{dC_t}{dC_0} = e^{\rho t}\frac{U_{C_0}}{U_{C_t}} \qquad (9.3)$$

where a subscript denotes the first partial derivative. Combining (9.1) and (9.3),

$$e^{rt} = e^{\rho t} \frac{U_{C_0}}{U_{C_t}} \tag{9.4}$$

Assuming that instantaneous utility is characterized by a constant relative rate of risk aversion η:

$$U = \frac{C^{1-\eta}}{1-\eta} \tag{9.5}$$

so that $U_C = C^{-\eta}$ and defining $C_t := e^{gt} C_0$, we have that

$$e^{rt} = e^{\rho t} \frac{C_0^{-\eta}}{e^{-\eta g t} C_0^{-\eta}} = e^{\rho t + \eta g t} \tag{9.6}$$

Equation (9.6) is known as the Ramsey rule, after Professor Frank Ramsey of Cambridge University, who published a variant of the above derivation in 1928.

9.2 Time discounting**

Climate change is a long-term, global, uncertain problem. If you do not care about the distant future, far-away lands, and remote probabilities, then climate change is of little concern.

The Ramsey rule says that the consumption rate of discount r consists of three components:

$$r = \rho + \eta g \tag{9.7}$$

First, we discount the future because we expect to be richer (g). Indeed, it would not make sense to transfer money from the relatively poor current self to a relatively rich future self. The faster the expected growth, the higher the discount rate. Second, the evaluation of "relatively poor" versus "relatively rich" depends on the curvature of the utility function, commonly referred to as the rate of risk aversion (η). Together η and g constitute the growth rate of utility. This part of the Ramsey rule is largely uncontroversial.

The third part is controversial. People discount the future for the sake of it being the future. This is captured by the pure rate of time preference (ρ), also referred to as the utility rate of discount. Philosophers and religious leaders have long maintained that we should not treat the future differently than the present, that is, $\rho = 0$. Yet people do. All empirical evidence suggests that people are impatient. The observed consumption discount rate typically exceeds the growth rate of consumption, corrected for the rate of risk aversion.

The Ramsey rule describes the consumption rate of discount from an individual perspective. If the economy is in a dynamic equilibrium, the consumption rate of discount equals the interest rate, the price of money on the capital market. If the interest rate were higher than the consumption rate of discount, the bank would pay more for your savings than you think they are worth. Savings would increase, and the price of capital (i.e., the interest rate) would fall as the supply of investment increases. Vice versa, if the interest rate were lower, savings would fall and the interest rate would rise.

9.3   Declining discount rates***

The standard discount factor, reflected in Equation (9.2), is exponential. This is because exponential (or geometric) discounting guarantees time consistency: The mere passage of time does not alter our decisions. This is easily seen. Suppose time s passes. Re-examining the decisions based on maximizing the welfare function in Equation (9.2), we multiply net present welfare by $e^{-\rho s}$. The first-order conditions for intertemporal trade-offs then become

$$U'_s e^{-\rho s} = U'_{s+t} e^{-\rho(s+t)} = U'_{s+t} e^{-\rho s} e^{-\rho t} \Leftrightarrow U'_s = U'_{s+t} e^{-\rho t} \qquad (9.8)$$

That is, $e^{-\rho s}$ which drops out of the equation; it does not influence behaviour. In other words, time consistency implies that intertemporal trade-offs are driven by the time passed between benefits and costs, rather than by the time at which benefits and costs are realized.

It is readily shown that exponential discounting implies time consistency. The reverse is also true: Time consistency implies exponential discounting. This is more difficult to establish, and omitted here.

Time consistency is a desirable characteristic. Exponential discounting has counterintuitive implications though. The relative difference between year 10 and year 11 is the same as the relative difference between year 100 and year 101 and between year 1000 and year 1001. As in Equation (9.8), s (the year of origin) is irrelevant; only t (the time passed) matters.

Intuitively, you may argue that the relative difference between year 10 and year 11 is more like the relative difference between year 100 and year 110,

Table 9.1 Discount factors and certainty equivalent discount rate

t	Value of $1000 after t years				Certainty equivalent discount rate
	1%	4%	7%	1% or 7%	
1	$990	$961	$932	$961	3.9%
10	$905	$670	$497	$701	3.1%
100	$368	$18	$1	$184	1.0%

and between year 1000 and year 1100. That intuition is supported by behavioural data, experimental data and survey data. Observed discount factors are hyperbolic rather than geometric. That is, as the time horizon expands, the discount rate appears to fall.

The discount rate also falls if we are uncertain about what discount rate to use, or if we disagree. This again follows from Equation (9.8) and is illustrated in Table 9.1. The first-order conditions are determined by the discount *factor* rather than the discount *rate*. Suppose we are uncertain about the discount rate. There is a 50% chance that it should be 1%, and a 50% chance that it should be 7%. Table 9.1 shows the corresponding discount factors. Obviously, benefits in the far future are worth little with a high discount rate; distant benefits are worth more with a low discount rate.

Table 9.1 also shows the discount factor for a discount rate of 4%, the average of 1% and 7%, and the average discount factor for a discount rate of 1% and 7%. The former is always smaller than the latter (by Jensen's inequality) and the gap widens as we look further into the future. This is further illustrated by inverting the average discount factor, which yields the certainty-equivalent discount rate. The certainty-equivalent discount rate is always smaller than the average discount rate and the gap widens as we look further into the future.

Table 9.1 gives a numerical illustration, but certainty-equivalent discount rate always falls over time. The intuition is as follows. Suppose that the 1% and 7% reflect disagreement, with one person defending one and someone else the other. Both have an opinion about intertemporal trade-offs in the short-run, and the average reflects both opinions. In the long-run, however, the person with the high discount rate is indifferent, but the person with the low discount rate is not; the latter's opinion therefore dominates.

The same reasoning holds for uncertainty about the discount rate. In one state of the world, you are indifferent about the long run; in another state of the world, you are not. The latter dominates. Therefore, the discount rate falls as we look further into the future.

None of this implies time inconsistency. Returning to the Ramsey rule (Equation (9.7)), the money discount rate may fall because we are uncertain about future economic growth. Time consistency requires that *utility* is discounted at a constant rate. Similarly, observed time preferences are not for utility; hyperbolic discounting for money is not inconsistent with exponential discounting for utility.

9.4 Axiomatic approaches to intertemporal welfare****

Equation (9.2) posits an intertemporal welfare function. This is a common approach – let us assume that, in this case, the discount factor declines exponentially. Often, such assumptions are justified with some reference to the plausible properties of the assumed function. The correct approach, however, is to first posit desirable properties (called axioms) and then derive the welfare function from those.

If the welfare function is used to assess welfare over an infinitely long time, it cannot simultaneously satisfy the axioms of anonymity and strong Pareto. Both axioms are desirable. The strong Pareto axiom is also referred to as Pareto superiority: Situation A is preferred to situation B if no one is worse off and at least one agent is better off.[1] This is obviously a desirable property.

Anonymity means that a welfare function should be indifferent to the question whether one agent or another gains an equivalent amount. This is obviously a desirable property, but it cannot be satisfied if strong Pareto is imposed too.

Intertemporal decisions are typically conceptualized as trade-offs between a sequence of generations. Generations arrive in a particular order. Anonymity is then a peculiar requirement. Dropping this, some function

$$W = \int_t U(C_t) D(t) \, dt \text{ with } \frac{\partial D}{\partial t} < 0 \qquad (9.9)$$

results. Note that the discount factor is some declining function of time, rather than the exponentially declining function used in Equation (9.2).

Equations (9.2) and (9.9) violate anonymity. Consider the following series of instantaneous welfare $\{1,2,1,1,1,\ldots\}$ and $\{1,1,2,1,1,\ldots\}$. Anonymity would imply indifference between the two situations. Discounted welfare would prefer the former over the latter.

Equations (9.2) and (9.9) do satisfy Strong Pareto:

$$\frac{\partial W}{\partial U(C_t)} > 0 \quad \forall t \qquad (9.10)$$

The axiomatic approach to intertemporal welfare goes back to the work of Tjalling Koopmans, a Nobel laureate in economics, in the 1960s. Although Koopmans was uneasy about discounting the future for the mere sake of it being the future – he clearly preferred $\rho=0$ in Equation (9.6) – his analysis shows that discounting is unavoidable.

Koopmans' result follows inescapably from his axioms. Other axioms would lead to different intertemporal welfare functions. Graciela Chichilnisky replaced the axiom of anonymity with axioms of independence and non-dictatorship. Francisco Alvarez-Cuadrado and Ngo Van Long dropped independence.

A net present welfare criterion violates the axiom of non-dictatorship of the present. Essentially, if we use Equation (9.9) to chart an optimal course for the entire future, then we let the preferences of the current generation dictate the policy choices of future generations. This is the way the world is, but perhaps not how it should be.

There is a welfare criterion that satisfies strong Pareto as well as axioms of non-dictatorship of the present and the poorest:

$$W = \vartheta \int_t U(C_t)D(t)\,dt + (1 - \vartheta)U(\underline{C}) \qquad (9.11)$$

That is, the welfare criterion – dubbed the Bentham–Rawls welfare criterion – is the weighted sum of standard net present welfare – the Bentham component – and the utility of the least-well-off generation – the Rawls component. If $\vartheta=1$, welfare is dictated by the concerns of the current generation; if $\vartheta=0$, the poorest generation dictates the outcome (up to the point where it no longer is the poorest). For $0 < \vartheta < 1$, neither the current nor the poorest generation dictates the result – unless the current generation is the poorest. This formulation puts a price on the current generation getting rich *at the expense* of future generations.

9.5   Equity**

The impact of climate change varies greatly between countries. Particularly, poor countries tend to be most vulnerable. The studies referred to in Chapter 6 estimate the impacts per country or region, and add up the results, in dollars, to the global total. This is wrong, as became clear in 1995 during the final stage of the preparation of the Second Assessment Report of Working Group 3 of the Intergovernmental Panel on Climate Change. Particularly, health risks are valued according to the national average willingness to pay. As the willingness to pay to reduce health risks is bound by the ability to pay (see Chapter 5), health risks in poor countries are deemed to be less severe than health risks in rich countries. In other words, we care more about the death of someone who is rich than about the death of someone who is poor.

Two solutions have been proposed to this moral conundrum. First, for a global problem such as climate change, global average values should be used. This does not work, however. If the Bangladeshi government would use the global average willingness to pay to reduce health risks for its climate policy, and the national average for its flood management, then it should shift its resources from flood defence to greenhouse gas emission reduction. If the UK government would use the global average for climate policy and the national average for flood management, then it should shift its resources from climate to floods. This cannot be the solution.

The other solution is to use equity weights. The problem with aggregating dollars is that a dollar to a poor woman is not the same as a dollar to a rich woman. Equity weights correct for this. Formally,

$$D_W = \sum_c \frac{1}{W_M} \frac{\partial W}{\partial U_c} \frac{\partial U_c}{\partial D_c} D_c \qquad (9.12)$$

The global damage D_W is a weighted sum of the country damages D_c. The weights have three components. First, the monetized damages D_c are transformed into utility by multiplying them with the marginal utility. Technically, marginal utility is the Jacobian that transforms from money space to utility space. The first partial derivative is like a ratio with utility as the unit in the nominator and money as the unit of the denominator. The unit of marginal utility is thus utils per dollar.

The second component of the equity weight is another Jacobian, transforming national utility into global welfare. The third component is a normalization. We transformed damages from money into utility, and utility into welfare. The global damages are expressed in money, however. The normalization thus has to be expressed in dollars per welfare. The term W_M in Equation (9.12) does exactly that. It considers the situation in which the global budget constraint is slightly loosened – say, a Martian comes along and gives us a dollar – and the windfall gain is optimally distributed over the world population. This normalization is appropriate in the sense that it is equity neutral.

If we assume that the utility function exhibits a constant relative rate of risk aversion, and that global welfare is the simple sum of national utilities – that is, a utilitarian welfare function in which we do not care about the distribution of utility, although we do care, through the curvature of the utility function, about the distribution on income – then Equation (10.8) becomes

$$D_W = \sum_c \left(\frac{y_W}{y_c}\right)^\eta D_c \qquad (9.13)$$

where y_c is the national average per capita income, y_W is the global average per capita income and η is the rate of risk aversion. Equation (9.13) is more intuitive. Damages in countries with a below average income receive a greater weight. The weight increases as the utility function is more curved. If utility is linear in income, $\eta = 0$, global damage is the simple sum of country damage. That is, the studies in Chapter 6 make the peculiar assumption of risk neutrality.

Returning to the justification, equity weights can imply global average values, for

$$D_W = \sum_c \left(\frac{y_W}{y_c}\right)^\eta D_c = \sum_c \left(\frac{y_W}{y_c}\right)^\eta V_c I_c = \sum_c \left(\frac{y_W}{y_c}\right)^\eta \left(\frac{y_c}{y_W}\right)^\varepsilon V_W I_c$$

$$= {}_{\eta=\varepsilon} \sum_c V_W I_c \qquad (9.14)$$

where I_c is the physical impact (say, the number of premature deaths), V_c is the national unit value (say, the willingness to pay to reduce health risks), V_W is the global average unit value, and ε is the income elasticity of that value. If $\eta = \varepsilon$, all impacts are valued at the global average. There is, of course, no reason why the rate of risk aversion would be equal to the income elasticity of willingness to pay.

9.6 Derivation of equity weights***

Above, equity weights are posited. The derivation is as follows. Assume a global, intertemporal welfare function

$$W = \sum_c \int_t U(C_c(t)) e^{-\rho t} dt \qquad (9.15)$$

where the global welfare W is the sum of the present welfare of a finite number of countries c. The present welfare of country c is the integral over the stream of utility U, derived from consumption C at time t, discounted at rate ρ.

The utility function is assumed to exhibit a constant relative rate of risk aversion, so

$$U = \frac{C^{1-\eta}}{1-\eta} \qquad (9.16)$$

where η is the rate of risk aversion.

The social cost of carbon is the effect of an infinitesimally small change in emissions on welfare

$$scc_w := \frac{\partial W}{\partial E(0)} = \sum_c \int_t \frac{\partial U(C_c(t))}{\partial C_c(t)} \frac{\partial C_c(t)}{\partial E(0)} e^{-\rho t} dt = \sum_c \int_t C_c(t)^{-\eta} \frac{\partial C_c(t)}{\partial E(0)} e^{-\rho t} dt$$

$$= \sum_c \int_t C_c(0)^{-\eta} e^{-\eta g t} \frac{\partial C_c(t)}{\partial E(0)} e^{-\rho t} dt = \sum_c C_c(0))^{-\eta} \int_t \frac{\partial C_c(t)}{\partial E(0)} e^{-(\rho + \eta g)t} dt$$

$$= \sum_c C_c(0)^{-\eta} SCC_c \qquad (9.17)$$

That is, the global social cost of carbon scc_w is the weighted sum of the national social costs of carbon scc_c.

Equation (9.17) does not specify the social cost of carbon as we normally think about it. Its unit is *utils* per tonne of carbon, rather than dollars per tonne of carbon. We therefore need to normalize it with something that is measured in utils per dollar. For a single agent, or a country with a representative agent, we would normalize by marginal utility with respect to consumption. Normalization for a global planner requires a bit more thought.

Consider the instantaneous global welfare function

$$W = \sum_c U(C_c) \tag{9.18}$$

Maximize this, subject to the budget constraint that total consumption cannot exceed total income M. Then the first-order conditions are

$$C_c^{-\eta} = \lambda \tag{9.19}$$

$$\sum_c C_c = M \tag{9.20}$$

The budget constraint is shared, so marginal utilities are equalized. The rate of risk aversion is assumed to be constant, so consumption is equalized. The solution is thus

$$\overline{C}^{-\eta} = \lambda \tag{9.21}$$

The shadow price of the budget constraint λ equals the increase in welfare if the budget constraint is slightly slackened. This is therefore an appropriate measure of marginal welfare with respect to consumption for a global planner.

Therefore, using (9.21) to normalize (9.17) we find

$$SCC_W = \frac{scc_W}{\overline{C}^{-\eta}} \sum_c \left(\frac{\overline{C}}{C_c} \right)^{-\eta} SCC_c \tag{9.22}$$

which is Equation (9.13).

9.7 Uncertainty**

Section 9.6 showed that care should be taken when aggregating dollar impacts over people with different incomes, and Section 9.2 argued the same for aggregation over time. Similar care should be taken when aggregating dollar impacts over different states of the world, each representing a possible but uncertain scenario of how the future might unfold.

Confronted with uncertainty about the impacts of climate change – itself due to uncertainty about emissions, about climate change, about vulnerability to climate change, about the impacts of climate change, and about the value of those impacts – it is tempting to calculate the expected impacts as follows:

$$ED = \sum_s p_s D_s \tag{9.23}$$

where E is the expectation operation, and p_s is the probability of the (discrete) state of the world s. If the uncertainty is continuous, Equation (9.23) is replaced by

$$ED = \int_D Df(D)\,dD \qquad (9.24)$$

Equations (9.23) and (9.24) are intuitive but wrong. The expected damage violates the Petersburg Paradox in that it assumes that large gains offset equally large losses of equal probability. In other words, the prospect of winning a million pounds with a 1% probability cancels the prospect of losing a million pounds with a 1% probability.

In Section 9.6, weights were introduced in aggregation. Essentially, we linearized the welfare problem. This would be inappropriate here because there are potentially large deviations in welfare. Therefore, the full welfare calculation is used:

$$EU(C, D) = \int_D U(C, D)f(D)\,dD \qquad (9.25)$$

The expected welfare loss is then

$$E\Delta U = U(C, D = 0) - EU(C, D) \qquad (9.26)$$

The certainty equivalent damage is defined by

$$U(C, D = 0) - U(C, D = CED) = U(C, D = 0) - EU(C, D) \qquad (9.27)$$

which can be reworked as

$$CED = U^{-1}[C, EU(C, D)] = U^{-1}\left[C, \int_D U(C, D)f(D)\,dD\right] \qquad (9.28)$$

That is, we compute the expected welfare loss and then invert the welfare function to obtain an impact measure in money. In (9.24), we first converted to money and then computed the expectation.

The risk premium is defined as

$$RPD = CED - ED \geq 0 \qquad (9.29)$$

The risk premium is positive. It is strictly positive for risk-averse actors (and if damages are indeed harmful).

Table 9.2 Illustration of the expected damage, the certainty equivalent damage, and the risk premium; italicized numbers are in utils, other numbers are in pound sterling

	impact	income		impact	income	
		100	1000		100	1000
0%	0	*4.61*	*6.91*		*4.61*	*6.91*
50%	10	*4.50*	*6.90*	10	*4.50*	*6.90*
50%	20	*4.38*	*6.89*	98	*0.69*	*6.80*
expectation	15	*4.44*	*6.89*	54	*2.60*	*6.85*
certainty equivalent		15.15	15.01		86.58	55.02
risk premium		0.15	0.01		32.58	1.02

Table 9.2 illustrates the difference between the expected damage and the certainty-equivalent damage. The risk premium is small if damages are small relative to income, but the risk premium rapidly grows if not.

9.8 Ambiguity***

Often, we are uncertain about what probability density function to use to describe the uncertainty about the parameter of interest. That is definitely the case in climate change. Different experts will give different opinions, not only about the most likely outcome, but also about the range of possible outcomes. You may be tempted to treat this as higher-order uncertainty, and define

$$EU(C, D) = \sum_s p_s \int_D U(C, D) f_s(D) \, dD \qquad (9.30)$$

where essentially we add another integration to the expected impacts, this time over all possible probability density functions f, weighted by their probability p_s.

However, Equation (9.30) violates the Allais Paradox: People prefer to enter a lottery with known probabilities over a lottery with uncertain probabilities – even if the convoluted probabilities are identical. This can be accommodated as follows

$$EV(C, D) = \sum_s p_s V \left[C, \int_D U(C, D) f_s(D) \, dD \right] \qquad (9.31)$$

The unambiguity equivalent damage is then defined as

$$AED = U^{-1} \{ V^{-1} [C, EV(C, D)] \} \qquad (9.32)$$

Inverse utility U^{-1} is needed to express the unambiguity equivalent in money; the term between the curly brackets is defined in the same way at the certainty equivalent. The unambiguity premium is defined as

$$APD = AED - CED \qquad (9.33)$$

If V is a linear function, $V(U) = U$, the unambiguity premium is zero. In other words, there is no ambiguity aversion and the unambiguity equivalent damage equals the certainty equivalent damage.

9.9 Deep uncertainty****

The expected value of the impact of climate is the integral over the impact times its probability; see Equation (9.25). In the tail of the distribution, the impact escalates but its chance falls precipitously. If the tail of the distribution is thin – and this is true in most cases – the probability falls faster than the damage grows. The product – chance times impact – thus falls to zero. The integral converges. The expectation exists and is bounded.

However, if the tail is fat, the probability does not fall as fast as we move out into the tail. Indeed, the damage grows faster than its chance falls. The product thus rises, and the integral does not converge. The expectation does not exist, or is infinitely large.

The Dismal Theorem, first published by Martin Weitzman of Harvard University, shows that the uncertainty about climate change and its impacts is such that tails are indeed fat.

Fat tails pose a problem. The expectation must exist if the aim is to maximize expected utility. Essentially, the Dismal Theorem has that expected utility maximization cannot be applied to climate policy – nor can cost–benefit analysis, as that is its monetized approximation.

Different people have interpreted the Dismal Theorem in different ways. One interpretation is that the Dismal Theorem formalizes the Precautionary Principle (in its 'it-is-better-to-be-safe-than-sorry' incarnation). A related interpretation is that stringent climate policy is justified. If the expected welfare loss is unbounded, then the social cost of carbon – its first partial derivative – is arbitrarily large. Therefore, optimal climate policy is arbitrarily stringent. This interpretation is incorrect. If expected utility does not exist, you cannot apply expected utility maximization to climate policy, and you cannot derive policy conclusions from its non-existence either. The only valid conclusion is that you need to find an alternative decision criterion other than expected utility.

Such alternatives are available, notably minimax regret. It works as follows. For each state of the world, find the optimal course of action – say, the carbon tax that maximizes welfare. Note that we do this *for each state of the world*, so that there is no uncertainty and fat tails are irrelevant. Define regret as the difference between the optimal welfare and actual welfare; in the optimum, regret is thus zero. Regret is defined for each state of the world. Then, across states of the world, find the maximum regret. Finally, across policies, find the intervention that minimizes maximum regret.

The main advantage of minimax regret is that, unlike expected welfare maximization, it can handle situations in which some policies have really bad consequences, perhaps only with a small change, and other policies do not. If all options lead to dreadful outcomes, minimax regret will select the least dreadful one.

Other objections to the Dismal Theorem have been raised too. In the original formulation, Weitzman only considers the impacts of climate change, which indeed may be very negative. Climate policy is a two-sided problem, however. The impact of emission abatement may be very negative too. Gradual emission reduction would not pose a serious problem to the economy (see Chapter 3), but in the short-run, fossil fuels are an essential input to the economy. In the long-run, we can improve energy efficiency and switch to alternative energy sources; in the short-run, reduced economic activity is the only option available to reduce emissions. Overly stringent climate policy thus bears a very substantial cost and this must be weighed against the risks of climate change.

Furthermore, the Dismal Theorem hangs on the welfare impact of very substantial climate change. In a logarithmic utility function (or any other CRRA one), welfare losses rapidly escalate if consumption falls below one (in whatever unit consumption happens to be measured). Other utility functions do not show this behaviour; utility falls with falling consumption, but not precipitously so at low levels of consumption. The Dismal Theorem therefore stands or falls with CRRA being a valid description of the behaviour of very poor people.

9.10 Implications for climate policy**

Figure 9.1 illustrates the implication for climate, focusing on the social cost of carbon.

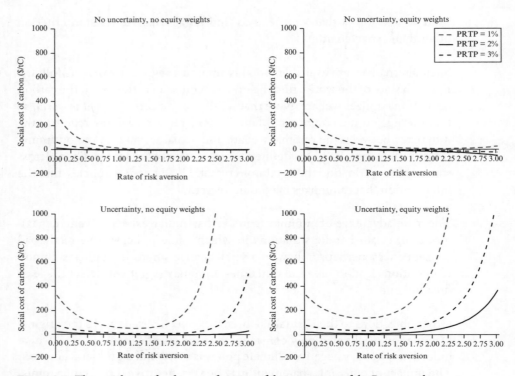

Figure 9.1 The social cost of carbon as a function of the parameters of the Ramsey rule

In Panel B, equity and uncertainty are ignored. The social cost of carbon is shown as a function of two parameters, viz. the pure rate of time preference and the rate of risk aversion. The relationships are simple. The higher the pure rate of time preference, the less you care about the future, and the lower the social cost of carbon. The rate of risk aversion only affects the discount rate (see the Ramsey rule, Equation (9.9)). The lower the rate of risk aversion, the lower the discount rate, the more you care about the future, and the higher the social cost of carbon.

Panel A introduces uncertainty (but ignores equity). It shows the certainty equivalent social cost of carbon as a function of the pure rate of time preference and the rate of risk aversion. The result is as expected. Because negative surprises are more likely than equiprobable positive surprises, the certainty equivalent social cost of carbon is larger than the best guess social cost of carbon. This effect becomes stronger as the rate of risk aversion increases.

Panel C introduces equity (but ignores uncertainty). For a pure rate of time preference of 1%, the results are intuitive albeit ambiguous. The lower the rate of risk aversion, the lower the discount rate, the more you care about the future, and the higher the social cost of carbon. At the same time, the lower the rate of risk aversion, the less you care about poor countries, and the lower the social cost of carbon. For higher pure rates of time preference, the relationship is more complex still. Because of carbon dioxide fertilization, the impacts of climate change on poor countries are positive in the short run (but negative in the long run). As the negative impacts in the long run are discounted away, the social cost of carbon becomes more negative as the rate of risk aversion increases.

Panel D uses both equity and uncertainty. The pattern is roughly the same as in Panel C, except that the uncertainty is such that the positive impacts in the best guess are more than offset by the negative impacts in the tails of the distribution. Most strikingly, the results span an enormous range. Depending on the choice of parameters, almost any carbon tax can be defended. This, of course, begs the question what parameters should be used.

9.11   The choice of parameters**

The pure rate of time preference and the rate of risk aversion are often referred to as "ethical parameters". For individuals, these parameters reflect the attitude towards the future and towards risk. Such attitudes may be based on moral reasoning or may result from social norms and upbringing. To an analyst, these parameters reflect the preferences of economic agents, and are measurable.

For society, these parameters also reflect attitudes, towards the future, towards risk, towards inequity within society, and towards inequity between societies. The parameters are measurable in the sense that decisions made can be interpreted as to their implied rates of time preference and risk aversion. However, what is and what ought to be are different things. Besides, analyses of the preferences revealed by government decisions show inconsistent behaviour. It is the role of decision analysts to improve policy making by removing inconsistencies and weeding out decisions that please no one. It is the role of moral leaders to improve preferences so that decisions are better, too, in the ethical sense of the word.

This is a deep issue. Reasonable people can and should reach different conclusions. The two polar positions are as follows. You can take a philosophical approach, and reason from first principles what the pure rate of time preference should be. If you find that difficult, you can take guidance from a thought leader such as Socrates, Jesus Christ, St Augustine, Mohammed,

Adolf Hitler, Johnny Rotten, Nick Stern, or Lady Gaga. Alternatively, you can argue that government decisions should reflect the will of the people and try to measure that will.

Both approaches are deeply flawed. Public policy by philosophical or religious principles is at risk of intolerance, authoritarianism and totalitarianism. The regimes that are deemed evil by history justified themselves by lofty principles and worthy aims.

Besides the potential errors and biases of measuring the will of the people, democracy is not the same as mob rule. A democratic government is supposed to safeguard minorities and use due process. A government is also supposed to provide public goods exactly because individuals cannot. In these cases, the collective will of the people deviates from the individual will of the people. It is a political question how much an elected government can and should deviate.

There is agreement too. Public policy should be internally consistent. If there is good reason to adopt a low pure rate of time preference for public investment in climate policy, then those reasons also apply to public investments in education, health care, and pensions. If there is good reason to worry about the impacts of climate change on people in other countries, then this should also affect aid, trade and migration policy.

9.12 Science and advocacy****

It should be clear to the reader by now that there is no such thing as best climate policy. Although the optimum is unambiguous and objective, it is conditional on a number of subjective choices, some of which are hotly disputed, while some other positions are widely but not universally supported. Any argument for a particular carbon tax is thus an argument for a particular social welfare function with a particular set of parameters. In fact, as shown in Chapter 4, any argument in favour of a carbon tax, of whatever level, rests on the assumption that concerns about economic efficiency can be traded off against concerns about climate efficacy.

In this sense, climate policy is no different from other policies that economists get involved in. Education, health, labour, and a range of other policies are based on a mix of positive and normative elements. For an individual researcher, it is important to distinguish between those parts of the analysis that are based on impartial interpretations of the available evidence, and those parts that are partial reflections on what society ought to do or be.

This is doubly important when speaking to lay people, who may not have the knowledge to draw the dividing line between what is and what ought to be as accurately as other experts would. An individual researcher must develop the understanding that her world views are not self-evidently true, probably not shared by everyone else and perhaps even repugnant to some. For a policy maker seeking advice, it is important to seek input from a number of experts with different perspectives.

To economists (and other social scientists) this comes almost naturally. The great economists of history – Smith, Ricardo, Mills, Marx, Keynes, Tinbergen, Friedman – were all deeply involved in the controversial policy issues of their times. The same is true for contemporary economists. It is made clear to young economists that they will not just study the economy, but help shape it. The example that was used during my induction was that one day, perhaps, one of us freshers would become governor of the central bank. Indeed, Ben Bernanke, one the greatest students of monetary policy, became Chairperson of the Federal Reserve, one of the most powerful makers of monetary policy.

Natural scientists do not generally share these sensitivities. Of the three core questions, natural scientists focus almost exclusively on "what if?" That is, they seek to develop an understanding of a particular aspect of the real world in the hope of gaining predictive skills. They aim to do so objectively, although they can only achieve replication. A group of culturally homogenous people would share the same blind spots; and the choice what to research and what not is of course a subjective one.

Social scientists tend to be more comfortable with the other two core questions, "so what?" – who would be hurt or helped if the predicted impacts come true and how big are these effects? – and "what to do?" – what is the appropriate course of action and its intensity to prevent, alleviate or stimulate the predicted impact? Natural scientists are often less comfortable with such questions.

Funtowicz and Ravetz even coined a new term – post-normal science – to describe research in areas where the policy stakes are high, the science is uncertain, and values are disputed. "Normal" refers to Kuhn, who was adamant that his description of the normal practice of research only refers to the natural sciences. "Post" suggests a chronological order in time, as if earlier debates about the solar system, the abolition of slavery, eugenics, and public pensions were not hugely controversial, both academically and politically. Post-normal science does make clear, however, that the conventional rules of natural science do not apply to a problem like climatic change. Unfortunately, post-normal science does not provide an alternative code of conduct (and is occasionally used to justify that anything goes).

The most important thing to remember is that low-quality research is irrelevant – or rather, that it should be. Flimsy results are often used to support a political position. That should not be. Policy should be informed by the best available knowledge.

Roger Pielke Jr provides a classification of the behaviour of experts in advising policy. The "pure scientist" may do policy-relevant research, but does not get involved in policy or policy advice, and her research agenda is set independently of policy concerns. The "science arbiter" restricts her role to predicting the impact of policies under discussion; she does not judge these impacts on their merits. The "issue advocate" seeks to restrict the number of policy options under consideration to the ones that adhere to her political convictions. The "honest broker", like the "science arbiter" assesses impacts of policy options, but also seeks to add new options to those already considered. Finally, the "stealth advocate" is actually an "issue advocate" but pretends to be a "science arbiter" or an "honest broker".

Obviously, one should strive to be an "honest broker". In the context of the discussion above, that means showing the sensitivity of key policy variables to parameters; and to discuss why some argue for one parameter value and others for a different number. I find it helps to state my personal opinion, but always make clear that I speak as a citizen rather than an expert, and add that others would disagree with me.

FURTHER READING

Christian Gollier's *The Economics of Risk and Time* (2004) and *Pricing the Planet's Future: The Economics of Discounting in an Uncertain Future* (2012) are excellent books on discounting, uncertainty, and the interactions between the two. Paul Portney and John Weyant's *Discounting and Intergenerational Equity* (1999) is a fine collection on discounting. Ferenc Toth's *Fair Weather: Equity Concerns in Climate Change* (2009) is one of the best treatises on equity and climate policy.
IDEAS/RePEc has a curated bibliography on this topic at http://biblio.repec.org/entry/tdg.html.

EXERCISES

9.1 Assume that the global mean surface air temperature (in deviation from pre-industrial times) rises from 0.8°C in 2000 to 3.8°C. Assume that the impact function is $-4.33T+1.92T^2$. Compute the stream of impacts for the 21st century. Discount these impacts back to 2000 using a 0%, 1%, 3%, 5%, and 10% consumption discount rate.

9.2 Aggregate the national impacts of climate change (at https://sites.google.com/site/climateconomics/data/06-economic-impacts) using equity weights with a rate of inequity aversion of 0, 1, 2 and 3.

9.3 Assume that the global mean surface air temperature (in deviation from pre-industrial times) rises from 0.8°C in 2000 to 3.8°C with a 50% probability, to 2.8°C with a 25% probability, and to 4.8°C with a 25% probability. Assume that the impact function is $-4.33T+1.92T^2$. Compute the stream of average impacts for the 21st century. How does this compare with the impact of a 3.8°C warming? Assume that per capita income rises from \$6,400/person/year in 2000 to \$86,000/person/year in 2100. Compute the certainty equivalent impact. How does this compare with the average impact? Assume that per capita income does not grow. How does this change the certainty equivalent impact?

9.4 The derivation of equity weights assumes a utilitarian welfare function. What would the equity weights be if welfare is given by

$$W = \sum_c \frac{U_c^{1-\gamma}}{1-\gamma}?$$ (9.34)

9.5 Take the survey on https://sites.google.com/site/climateconomics/data/14-climate-policy. The results can be downloaded at the bottom of the page. Compute the risk and ambiguity aversion from Questions 1 and 2, the discount rate from Questions 3 and 4, and the inequity aversion from Questions 5 and 6.

9.6 Read and discuss:

- **D. Anthoff, C. Hepburn and R.S.J. Tol (2009), 'Equity weighing and the marginal, damage costs of climate change', *Ecological Economics*, **68**, 836–849.
- **D. Anthoff and R.S.J. Tol (2010), 'On international equity weights and national decision making on climate change', *Journal of Environmental Economics and Management*, **60**, 14–20.
- ***M.L. Weitzman (2009), 'On modeling and interpretating the economics of catastrophic climate change', *Review of Economics and Statistics*, **91** (1), 1–19.
- ***W.D. Nordhaus (2011), 'The economics of tail events with an application to climate change', *Review of Environmental Economics and Policy*, **5** (2), 240–257.
- ****A. Millner (2013), 'On welfare frameworks and catastrophic climate risks', *Journal of Environmental Economics and Management*, **65**, 310–325.
- ****S. Dietz and G.B. Asheim (2012), 'Climate policy under sustainable discounted utilitarianism', *Journal of Environmental Economics and Management*, **63**, 321–335.
- ****R.S.J. Tol (2013), 'Climate policy with Bentham–Rawls preferences', *Economics Letters*, **118**, 424–428.
- ****A. Millner, S. Dietz and G.M. Heal (2013), 'Scientific ambiguity and climate policy', *Environmental and Resource Economics*, **55**, 21–46.
- ****M. Ha Duong and N. Treich (2004), 'Risk aversion, intergenerational equity and climate change', *Environmental and Resource Economics*, **28**, 195–207.

NOTE

1 In weak Pareto, all need to be better off.

10

Irreversibility and learning

TWEET BOOK

- Climate policy is more stringent under uncertainty because negative surprises are more likely than positive surprises. #climateeconomics
- Climate policy is more stringent under uncertainty because people are risk averse. #climateeconomics
- Irreversibility combined with uncertainty calls for yet more caution: Mistakes cannot be undone. #climateeconomics
- Greenhouse gas emissions stay in the atmosphere for a long time and lock us into climate change. #climateeconomics
- Energy and transport capital are long-lived and lock us into levels and patterns of energy use. #climateeconomics
- The risk of being locked into undesirable climate change is greater than the risk of being locked into expensive energy. #climateeconomics
- Irreversibility thus calls for more stringent climate policy. #climateeconomics
- Optimal emission reduction today depends on expected climate policy in the future. #climateeconomics
- Future climate policy depends on future knowledge. The prospect of learning thus affects current climate policy. #climateeconomics
- Future learning implies that current climate policy needs to hedge less against future policy mistakes. #climateeconomics
- Therefore, the prospect of learning in the future implies that optimal climate policy is less stringent today. #climateeconomics

10.1 Introduction

Section 8.3 introduced benefit–cost analysis for both static and dynamic problems. Section 9.8 discussed benefit–cost analysis under risk. In a risk context, parameters are not known with certainty but their probability density functions are. Here, we take both issues a step further, introducing irreversibilities and learning.

The impact of irreversibility on decisions is intuitive:

- If I would do something fun today that will cause a disaster the day after tomorrow, then I should not do it.
- If I would do something fun today that might cause a disaster the day after tomorrow, then I should wonder whether it is worth the risk and probably not do it.
- If I would do something fun today that might cause a disaster the day after tomorrow, but I can undo my actions without cost tomorrow when I will know more, then I should do it.
- If I would do something fun today that might cause a disaster the day after tomorrow, but I can undo my actions with a cost tomorrow when I will know more, then I should wonder whether it is worth the risk.

In other words, actions with irreversible consequences should be considered more carefully than actions with reversible consequences.

The impact of learning is not trivial. It is obvious that as we learn more about the climate problem, we can refine our actions. We do not know, however, what we will learn – if we would, we would have learned it already. The fact that we know that we will learn affects current optimal policy, even if we do not know what we will learn. This is counterintuitive. The rest of the chapter explains how.

10.2 A stylized example**

Consider a three period problem. Greenhouse gases are emitted in periods 1 and 2, accumulate in the atmosphere, and do damage in period 3. We seek to minimize the net present value of the sum of the abatement costs and the damage costs, by setting emission reduction targets in periods 1 and 2.

We will consider four variants of this problem. In every variant, uncontrolled emissions E are 100 units in period 1 and 2. The atmospheric concentration equals 90% of the concentration in the previous period plus the emissions in the previous period. Without emission control, the atmospheric concentration is therefore 190 units. Emission reduction costs C are quadratic in emission reduction effort R, with unit marginal costs, that is:

$$C_t = 0.5R_t^2 \Rightarrow \frac{\partial C_t}{\partial R_t} = R_t t = 1,2 \qquad (10.1)$$

Controlled emissions equal $E_t - R_t$. The discount rate is 10%. All parameters in the model are known with certainty, except for the damages of climate change: Damages are either high or low, with a 50% chance. In the second and fourth variant of the model, it is revealed in the second period whether damages are high or low. In the first and third variant, there is no learning. In the first two variants, impacts are linear in concentration. In the last two variants, impacts are quadratic

	No learning	Learning
Linear impacts	Variant 1	Variant 2
Quadratic impacts	Variant 3	Variant 4

In the first two variants, damage D is 10 times the atmospheric concentration in period 3 with a 50% probability, and 20 times that with a 50% chance. Without learning, the problem is then

$$\min_{R_1,R_2} NPV = 0.5R_1^2 + \frac{0.5R_2^2}{1+0.1} + (0.5 \cdot 10 + 0.5 \cdot 20)\frac{0.9(100-R_1)+100-R_2}{(1+0.1)^2} \quad (10.2)$$

This is an unconstrained maximization – because we substituted the stock function into the objective function – so the first-order conditions for optimality are

$$\frac{\partial NPV}{\partial R_1} = 0 \Leftrightarrow R_1 = \frac{(0.5 \cdot 10 + 0.5 \cdot 20)0.9}{(1+0.1)^2} = 10.9 \quad (10.3)$$

$$\frac{\partial NPV}{\partial R_2} = 0 \Leftrightarrow R_2 = \frac{(0.5 \cdot 10 + 0.5 \cdot 20)}{1+0.1} = 13.5 \quad (10.4)$$

With learning, we cannot use Equation (10.2) because decisions are made conditional on different information. It is common to solve problems like these by backward induction, first solving the optimization problem of the final period and then working towards the present. That is, for period 2

$$\min_{R_2} NPV_2 = 0.5R_2^2 + \alpha\frac{0.9(100 - R_1) + 100 - R_2}{1+0.1} \alpha = 10, 20 \quad (10.5)$$

The first-order conditions are

$$\frac{\partial NPV_2}{\partial R_2} = 0 \Leftrightarrow R_{2,i} = \frac{a}{1+0.1} = 9,18 \quad (10.6)$$

Without learning, the optimal decision is to reduce emissions in period 2 by 13.5 units. With learning, the optimal decision is to reduce emissions by either 9 or 18 units. On average, the decision is the same – $(9+18)/2=27/2=13.5$ – but in reality it is not.

For period 1, the problem is

$$\min_{R_1} NPV_1 = 0.5R_1^2 + (0.5 \cdot 10 + 0.5 \cdot 20)\frac{0.9(100 - R_1) + 100 - (0.5 \cdot R_{2,1} + 0.5 \cdot R_{2,2})}{(1 + 0.1)^2} \tag{10.7}$$

That is, the optimal action in the first period depends on the *expected* action in the second period. The first-order conditions are

$$\frac{\partial NPV_1}{\partial R_1} = 0 \Leftrightarrow R_1 = \frac{(0.5 \cdot 10 + 0.5 \cdot 20)0.9}{(1 + 0.1)^2} = 10.9 \tag{10.8}$$

Learning does not affect the optimal decision in the first period. There is no difference between (10.3) and (10.8): Because optimal emission reduction in period 2 is proportional to impacts, the average emission reduction equals the emission reduction at the average impact. More formally, the expectation of the maximum (exp max) equals the maximum of the expectation (max exp).

In general, exp max ≠ max exp. Both the expectation (exp) and the maximization (max) are so-called operators. You can only switch the order of operators if all are linear. The expectation is a linear operator; the mean is a weighted sum. The maximization is not a linear operator – unless the first-order conditions are linear. They are in this case.

In the third and fourth variant, damage D is proportional to the atmospheric concentration squared; the parameter is 0.026 with a 50% probability, and 0.053 with 50% chance; these parameters are chosen such that the marginal damages are equal in the no-control case to the ones above. Without learning, the problem is then

$$\min_{R_1, R_2} NPV = 0.5R_1^2 + \frac{0.5R_2^2}{1 + 0.1} + (0.5 \cdot 10 + 0.5 \cdot 20)\frac{(0.9(100 - R_1) + 100 - R_2)^2}{(1 + 0.1)^2} \tag{10.9}$$

The first-order conditions are

$$\frac{\partial NPV}{\partial R_1} = R_1 - (0.5 \cdot 10 + 0.5 \cdot 20)\frac{2 \cdot 0.9(0.9(100 - R_1) + 100 - R_2)}{(1 + 0.1)^2} = 0 \Leftrightarrow \tag{10.10}$$

$$R_1 = \frac{(0.5 \cdot 10 + 0.5 \cdot 20)2 \cdot 0.9(0.9 \cdot 100 + 100 - R_2)}{(1 + 0.1)^2 + 2 \cdot 0.9 \cdot 0.9(0.5 \cdot 10 + 0.5 \cdot 20)}$$

$$\frac{\partial NPV}{\partial R_2} = \frac{R_2}{1+0.1} - (0.5 \cdot 10 + 0.5 \cdot 20)\frac{2(0.9(100-R_1)+100-R_2)}{(1+0.1)^2} = 0 \Leftrightarrow \tag{10.11}$$

$$R_2 = \frac{(0.5 \cdot 10 + 0.5 \cdot 20)2(0.9(100 - R_1) + 100)}{(1 + 0.1) + 2(0.5 \cdot 10 + 0.5 \cdot 20)}$$

This is a system of two linear equations with two unknowns. It solves as R_1=9.92 and R_2=12.13. The crucial difference with the first variant is that the optimal decisions in the two periods interact: R_1 depends on R_2 and R_2 depends on R_1.

With learning, the first order condition for period two equals

$$\frac{\partial NPV}{\partial R_2} = \frac{R_2}{1 + 0.1} - \alpha\frac{2(0.9(100 - R_1) + 100 - R_2)}{(1 + 0.1)^2} = 0 \Leftrightarrow \tag{10.12}$$

$$R_2 = \frac{2\alpha(0.9(100 - R_1) + 100)}{(1 + 0.1) + 2\alpha}$$

For period one, this is

$$\frac{\partial NPV}{\partial R_1} = R_1 - (0.5 \cdot 10 + 0.5 \cdot 20)\frac{2 \cdot 0.9(0.9(100-R_1)+100-0.5R_{2,1}-0.5R_{2,2})}{(1+0.1)^2} = 0 \Leftrightarrow \tag{10.13}$$

$$R_1 = \frac{(0.5 \cdot 10 + 0.5 \cdot 20)2 \cdot 0.9(0.9 \cdot 100 + 100 - 0.5R_{2,1} - 0.5R_{2,2})}{(1 + 0.1)^2 + 2 \cdot 0.9 \cdot 0.9(0.5 \cdot 10 + 0.5 \cdot 20)}$$

The difference between (10.10) and (10.13) is that the former has the emission reduction in the second period based on the average damage, whereas the latter has the average emission reduction in the second period.

With learning, the solution is R_2=8.72, 15.82 with an average of 12.05. This compares to R_2=12.13 without learning. The reason is intuitive. Although the problem has been set-up with linear utility (or zero risk aversion), the damage function is curved. This acts like risk aversion. Therefore, if the damages are unknown, a rational decision maker would err on the safe side, and be closer to the high damage scenarios.

With learning, R_1=9.86 compared to R_1=9.92 without learning. The intuition is as follows. With learning, underinvestment in emission reduction will be recognized and can be corrected in period 2. Without learning, a rational decision maker is more cautious. She needs to contend not only with uncertainty about the state of the world, but also with an imperfectly informed successor.

10.3 Applications to climate change**

The previous section argued, using a stylized example, that in an imperfectly known, dynamic, non-linear system with irreversibilities, the prospect of future learning affects current optimal behaviour – even if it is not known what will be learned. The intuition is as follows. Net present welfare depends on future choices. Therefore, current choices are influenced by future choices. If there is learning, future choice will be different than if there is no learning. Therefore, future learning affects current choices.

The climate problem obviously meets the criteria: uncertainty, dynamics, non-linearity and irreversibility characterize the costs and benefits of greenhouse gas emission reduction. The time horizon is long enough that learning is inevitable. Therefore, from the analysis above, we would expect that the prospect of future learning affects optimal greenhouse gas emission reduction.

Climate policy is far more complicated than the stylized model above. A crucial difference is that there are irreversibilities on both sides of the equation. Carbon dioxide remains in the atmosphere for a long time. On the other hand, capital is long-lived. An investment in renewable energy cannot be reversed without accepting the cost of capital destruction. Optimal emission reduction thus balances two irreversibilities, and learning pushes emission control in both directions at once.

Figure 10.1 shows the impact of future learning on optimal emission reduction in the near-term, measured as the percentage change from optimal emission reduction without learning. There are 17 estimates in Figure 10.1, taken from five studies (each reporting results for a few sensitivity analyses). One estimate has that the irreversibility of emission abatement outweighs the irreversibility of emissions; optimal emissions are thus lower (or optimal abatement more stringent) due to learning. Two estimates are indistinguishable from zero. Fourteen estimates have that learning increases optimal emissions (or makes optimal abatement less lenient). The estimated effect spans two orders of magnitude, from 0.6% to 63%, with most estimates in the high teens. That is, learning is more than an intellectual curiosity. Learning appears to have a substantial effect on optimal emission abatement. Specifically, because

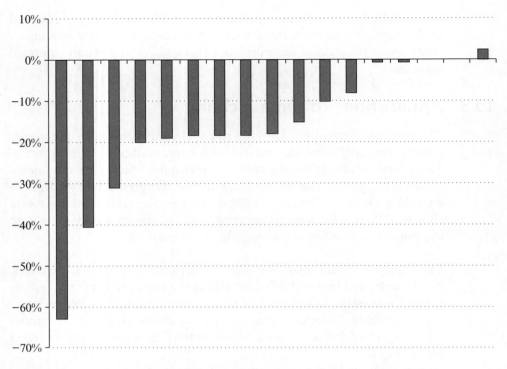

Figure 10.1 The effect of future learning on near-term optimal emission reduction (relative to emission reduction without learning) according to different studies and model parameterizations

we can reasonably expect to know more in the future, we can afford a more lenient climate policy today.

FURTHER READING

The best introduction to uncertainty, irreversibility and learning is *Buying Greenhouse Insurance* (1991) by Alan Manne and Richard Richels. Highlights were published in 1991 in *Energy Policy* under the same title. Ingham, Ma and Ulph's 'Climate change, mitigation and adaptation with uncertainty and learning', *Energy Policy* (2007) has a good review of the recent literature.

IDEAS/RePEc has a curated bibliography on this topic at http://biblio.repec.org/entry/tdh.html.

EXERCISES

10.1 Derive Equations (10.1)-(10.13) for a 50% probability of an impact of 9 and a 50% probability of an impact of 21. Repeat again for a 50% probability of an impact of 11 and a 50% probability of an impact of 21. What is the relative impact of a mean-preserving increase in spread (first exercise) versus a spread-preserving increase in mean (second exercise)?

10.2 Derive Equations (10.1)-(10.13) if 100% of the carbon dioxide stays in the atmosphere for the next period, and if 0% does. What is the impact on the effect of learning on optimal control?

10.3 Read and discuss:

- **C.D. Kolstad (1994), 'George Bush v Al Gore: Irreversibilities in greenhouse gas accumulation and emission control investment', *Energy Policy*, **22** (9), 771–778.
- **A.S. Manne and R.G. Richels (1995), 'Greenhouse debate: Economic efficiency, burden sharing, and hedging strategies', *Energy Journal*, **16** (4), 1–37.
- ***R.S. Pindyck (2007), 'Uncertainty in environmental economics', *Review of Environmental Economics and Policy*, **1** (1), 45–65.
- ***K. Keller, B.M. Bolker and D.F. Bradford (2004), 'Uncertain climate thresholds and optimal economic growth', *Journal of Environmental Economics and Management*, **48** (1), 723–741.
- ****A. Ingham, J. Ma and A.M. Ulph (2007), 'Climate change, mitigation and adaptation with uncertainty and learning', *Energy Policy*, **35**, 5354–5369.
- ****A. Lange and N. Treich (2008), 'Uncertainty, learning and ambiguity in economic models on climate policy: Some classical results and new directions', *Climatic Change*, **89**, 7–21.
- ****E. Baker (2005), 'Uncertainty and learning in a strategic environment: Global climate change', *Resource and Energy Economics*, **27**, 19–40.

11

International environmental agreements

11.1 Cooperative and non-cooperative abatement**

Chapter 8 discusses climate policy from the perspective of a global social planner. This is a useful yardstick. A global social planner can maximize global welfare. This is the best climate policy. At the same time, this is an unrealistic perspective. There is nothing that remotely resembles a global social planner, in that no institution can force emission reduction policies on sovereign countries let alone mobilize the transfers needed to turn a potential Pareto improvement into an actual one. More realistic representations of climate policy therefore must lead to lower welfare than in global optimum.

Figure 11.1 shows a regional break-down of the social cost of carbon. In this particular example, the global social cost of carbon is $16/tC. The global social cost of carbon is the sum of the regional social costs of carbon, which are by definition a fraction of the global cost. If the world were run by 16 regional social planners who ignore their impact on the rest of the world,

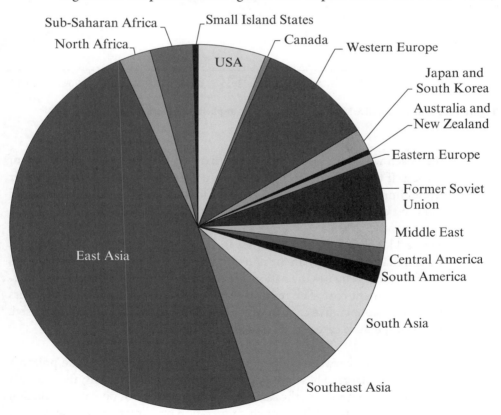

Figure 11.1 Regional breakdown of the social cost of carbon

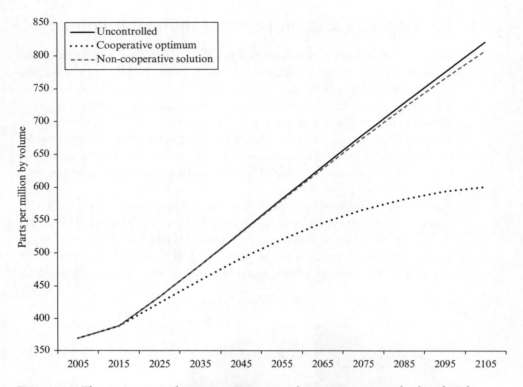

Figure 11.2 The cooperative and non-cooperative atmospheric concentration of carbon dioxide

then each would impose a carbon tax that equals the regional social cost of carbon. If the numbers of Figure 11.1 are correct, the global social planner would impose a carbon tax of \$16/tC. The European social planner would impose a carbon tax of almost 10% of \$16/tC, that is, ¢1.5/tC.

Figure 11.2 shows the implications. In Figure 11.2, there are 190 countries (rather than 16 regions). In the non-cooperative scenario, each of the national social planners equates the national social cost of carbon to the national marginal abatement cost. The result is some emission reduction, but much less than in the cooperative case, in which the global social cost of carbon is used. (The cooperative case was discussed in Chapter 8.)

The difference between cooperative and non-cooperative climate policy is intuitive and large. These are polar cases, however. Obviously, the world is not run cooperatively. At the same time, countries do not operate in isolation either. The question is how much cooperation on climate policy can be sustained by sovereign, self-interested nations.

11.2 Free-riding**

Free-riding is a customary if somewhat peculiar way to study the provision of public goods. Full cooperation is the starting point of the analysis. Then, each country considers whether it wants to continue to cooperate. Countries do not take other countries' incentives to cooperate into account. This is the best response in the Nash sense of the word, and therefore this disregard for the plans of others is often called Nash behaviour. Nash is not the originator of this assumption, however, and a more accurate description would be to call this myopic behaviour. That is, a country considers whether it wants to continue to cooperate, assuming that the other countries do continue to cooperate. The trade-off then is between the cost savings due to lower emission reduction made versus the additional damages incurred.

Let us investigate this a bit more using a linear-quadratic game. The costs C of emission reduction R for country i are

$$C_i = \alpha_i R_i^2 \qquad (11.1)$$

where α is a parameter, denoting the unit cost of emission reduction. By assumption, the costs of emission reduction of country i only depend on emission reduction in country i. Emission reduction in other countries has no effect.

The benefits B of emission reduction are

$$B_i = \beta_i \sum_j R_j E_j \qquad (11.2)$$

where β is a parameter, denoting the social cost of carbon, and E are emissions in the absence of climate policy. That is, the benefits of the emission reduction depend on the actions of all countries, weighted by their emission reduction effort R and their initial emissions E.

In the non-cooperative solution, each country maximizes its own net benefits by equating the marginal cost and benefits:

$$\frac{\partial B_i}{\partial R_i} = \beta_i E_i = 2\alpha_i R_i = \frac{\partial C_i}{\partial R_i} \forall i \Rightarrow R_i' = \frac{\beta_i E_i}{2\alpha_i} \forall i \qquad (11.3)$$

In the cooperative solution, all countries jointly maximize their collective net benefits by equating the marginal costs and benefits:

$$\frac{\partial \sum\limits_{j} B_j}{\partial R_i} = E_i \sum_{j} \beta_j = 2\alpha_i R_i = \frac{\partial C_i}{\partial R_i} \forall i \Rightarrow R_i' = \frac{\sum\limits_{j} \beta_j E_i}{2\alpha_i} = :\frac{\beta E_i}{2\alpha_i} \forall i \ (11.4)$$

At first sight, the cooperative and non-cooperative solutions look very similar. In both cases, optimal emission reduction equals the marginal benefits times own emissions over two times the unit abatement cost. There is one crucial difference, however. In the non-cooperative solution, only the marginal benefits to the own country are considered, whereas in the cooperative solution the marginal benefits to all countries are considered.

The difference in costs follows from substituting (9.3) and (9.4) into (9.1):

$$C_i' - C_i^* = \alpha_i \left(\frac{\beta E_i}{2\alpha_i}\right)^2 - \alpha_i \left(\frac{\beta_i E_i}{2\alpha_i}\right)^2 = (\beta^2 - \beta_i^2) \frac{E_i^2}{4\alpha_i} \forall i \quad (11.5)$$

As we are considering free-riding, the difference in benefits follows from the change in emission reduction by the own country only:

$$B_i' - B_i^* = \beta_i \left(\frac{\beta E_i}{2\alpha_i} - \frac{\beta_i E_i}{2\alpha_i}\right) E_i = (\beta_i - \beta_i^2) \frac{E_i^2}{2\alpha_i} \forall i \quad (11.6)$$

It would be in a country's best interest to free-riding if the cost savings of Equation (11.5) exceed the additional damages of Equation (11.6).

This is almost always the case. Table 11.1 illustrates this. For convenience, emissions are set equal to one, and unit costs to one-half. In the first row, the national social cost of carbon is one-quarter of the global social cost. Then, the cost-savings are 7.5 and the additional damages are 3. It is better to free-ride. In the second row, the national social cost of carbon is one-tenth of the global social cost. Then, the cost-savings are higher: 49.5. They are much higher as a smaller country would need to do much more for the rest of world. The additional benefits are higher too: 9. But, since the emission reduction costs are quadratic and the emission reduction

Table 11.1 Free-riding illustrated

β	β_i	A	E	ΔC	ΔB	$\Delta C - \Delta B$
4	1	0.5	1	7.5	3	3.5
10	1	0.5	1	49.5	9	40.5
2	1	0.5	1	1.5	1	0.5

benefits linear, also in this case it is better to free-ride. In the third row, the national social cost of carbon is one-half of the global social cost. The cost-savings fall to 1.5. The additional damages fall too, to 1. It is still better to free-ride.

In this set-up, every country has an incentive to free-ride. In a more general set-up, almost every country would have. As a result, cooperation collapses.

11.3 Cartel formation**

Above, we contrast full cooperation and non-cooperation. There are intermediate cases too, in which some countries cooperate with one another and others do not. This is usually studied with the help of cartel formation games (which, as the name suggests, originated in industrial organization).

A coalition is said to be stable if and only if it is internally stable, externally stable, and profitable. A coalition is said to be internally stable if none of its members is better off outside the coalition. A coalition is said to be externally stable if none of its non-members is better off inside the coalition. This is intuitive: A coalition is stable if no one wants to leave and no one wants to join. A coalition is said to be profitable if all of its members are at least as well off as in the fully non-cooperative case. This final condition partly overcomes the myopic nature of the first two conditions (which only consider a single move by a single agent). An alternative interpretation is that cartel theory takes the non-cooperative solution as its starting point.

The grand coalition (which contains all agents) is always externally stable (as there are no non-members). For a coalition of two, internal stability and profitability are the same.

Cartel formation can be illustrated with the linear-quadratic game of Section 11.2. For simplicity, assume that there are two agents only. The costs of emission reduction are then

$$C_1 = \alpha_1 R_1^2; \; C_2 = \alpha_2 R_2^2 \tag{11.7}$$

The benefits of emission reduction are

$$B_1 = \beta_1(R_1 E_1 + R_2 E_2); \; B_2 = \beta_2(R_1 E_1 + R_2 E_2) \tag{11.8}$$

The non-cooperative solution is

$$R_1^* = \frac{\beta_1 E_1}{2\alpha_1}; \; R_2^* = \frac{\beta_2 E_2}{2\alpha_2} \qquad (11.9)$$

The cooperative solution is

$$R_1' = \frac{(\beta_1 + \beta_2)E_1}{2\alpha_1}; \; R_2' = \frac{(\beta_1 + \beta_2)E_2}{2\alpha_2} \qquad (11.10)$$

Subtracting (9.8) from (9.9), we find that the extra emission reduction in the cooperative case equals

$$\Delta R_1 = \frac{\beta_2 E_1}{2\alpha_1}; \; \Delta R_2 = \frac{\beta_1 E_2}{2\alpha_2} \qquad (11.11)$$

Cooperation can thus be interpreted as barter trade. Player 1 reduces his emissions further (at a cost to player 1) and in return player 2 further reduces her emissions as well (which benefits player 1). Cooperation is in player 1's interest if the benefits exceed the costs. Player 1 would like to increase his abatement by a little bit and get a lot of additional abatement by player 2 in return. However, player 2 wants the same: Do a little more herself, and hope that the other offers a lot in return. A deal can only be made if both parties are better off. A stable cooperation thus meets the criterion

$$\Delta C_1 < \Delta B_1 \; \wedge \; \Delta C_2 < \Delta B_2 \qquad (11.12)$$

For a coalition of N players, this becomes

$$\Delta C_1 < \Delta B_1 \; \wedge \; \Delta C_2 < \Delta B_2 \; \wedge \; \ldots \; \wedge \; \Delta C_N < \Delta B_N \quad (11.13)$$

This is a stringent set of conditions. As more players are added to the coalition, each coalition member is asked to do more at accelerating cost. The benefits increase too, but the benefits are reaped by members and non-members alike. Equation (11.12) thus implies one solution to the cartel formation game: Stable international environmental agreements have few signatories.

There is, in fact, another solution. There may be many signatories, each committing to little over and above what they would have done anyway. That is, international environmental agreements are either wide and shallow (many signatories not doing much) or deep and narrow (few signatories doing a lot). In either case, the impact on global emissions is limited.

11.4 Multiple coalitions****

Above, we use cartel theory to study the formation of a single coalition. A coalition is d'Aspremont stable if it is profitable, if no one wants to leave, and if no one wants to join. We can use a similar set-up for multiple coalitions: A set of coalitions is stable if each coalition is profitable, if no coalition member wants to leave to play non-cooperatively, if no non-cooperative players want to join a coalition, *and if no coalition member wants to switch to a different coalition.*

On the one hand, multiple coalitions allow for more choice. You would therefore expect welfare to improve. On the other hand, multiple coalitions impose an additional constraint on the equilibrium, namely inter-coalition stability. There is no such condition for one coalition, two for two coalitions, six for three coalitions, and $N(N-1)$ for N coalitions. This implies the solution space rapidly shrinks as more coalitions are added. In other words, multiple coalitions cannot reduce emissions by much more than can a single coalition.

What about two coalitions? Coalitions are formed simultaneously, but intuition is clearer for sequential coalition formation. Suppose there are a large number of players. The first coalition forms. Stability dictates that the coalition is small. There are therefore a large number of non-cooperative players left. A second coalition would thus form under conditions that are very similar as when the first coalition formed. The second coalition is therefore small too.

The same would be true for the third coalition, and the number of inter-coalition conditions is six (three times two) so that the space of feasible solutions is small. With four coalitions, there are twelve inter-coalition conditions (on top of four internal stability, four external stability, and four profitability conditions), so that solutions are likely to be infeasible.

Now suppose there are a small number of players. The first coalition forms. The impact of the formation of that first coalition on the remaining, non-cooperative players is relatively large. A second coalition can thus make a difference. If there are few players to begin with, two coalitions can have a large fraction playing cooperatively.

This is a paradoxical result: Multiple coalitions matter more for negotiations with few players than for many players.

11.5 International climate policy**

International climate policy has a long history of good intentions but few successes.

Anthropogenic climate change was put on the academic agenda in 1896 by Svante Arrhenius, winner of the 1903 Nobel Prize for Chemistry. His analysis was textbook material within decades, but although many meteorologists were employed by the state, it was not until 1988 that climate change was put on the political agenda. Four years later, the United Nations Framework Convention on Climate Change (UNFCCC) was negotiated in Rio de Janeiro.

The UNFCCC entered into force in 1994. It has been ratified by all UN members (except South Sudan) and a few non-UN entities as well (notably the European Union). This is no surprise, as the UNFCCC does not contain many commitments. Four are worth mentioning. The UNFCCC sets up an international system to standardize and report measurements of greenhouse gas emissions. This is important because it permits international comparison of data and performance. The UNFCCC calls for a stabilization of the atmospheric concentration of greenhouse gas, which implies 100% emission reduction for carbon dioxide (cf. Chapter 8). The UNFCCC further establishes that the responsibilities for climate policy are "common but differentiated" which is typically seen to indicate that rich countries should take the lead. Finally, the UNFCCC commits countries to negotiate. There is one big conference per year. In recent years, the number of participants was measured in the tens of thousands. Over the years, the number of smaller, preparatory, intermediate, committee and subcommittee meetings has grown steadily so that there are now civil servants who are employed full-time on the international climate negotiations.

These UNFCCC conferences aim to create international climate policy. Countries were close to a breakthrough in 1995 in Berlin. A deal was done in 1997 in Kyoto. The Kyoto Protocol establishes two things. First, the Kyoto Protocol sets up a (widely used) system through which rich countries can invest in greenhouse gas emission reduction in other, poorer countries; and a (rarely used) system through which rich countries can internationally trade emission permits. Second, the Kyoto Protocol defines emission targets for rich countries for the period 2008–12. Unfortunately, the Kyoto Protocol puts an undefined limit on the use of international flexibility mechanisms, it does not define emissions, and it does not specify sanctions for failing to meet the targets.

These issues were revisited in The Hague in 2000, with vice-president Gore eager to do a deal in support of his bid for the presidency. However, the countries of the European Union so vigorously disagreed with one another that the meeting collapsed. Since then, the EU has agreed on a common position well in advance of the international negotiations. As this position is public, the EU has de facto withdrawn from the negotiations: The EU position is known and immutable. Other countries do not need to talk to the EU as they can read the EU response in advance.

Shortly after the meeting in The Hague, George W. Bush was elected 43rd President of the USA. Although Bush had campaigned with a promise of a tax on greenhouse gas emissions, one of his first acts in office was to pull out of the Kyoto Protocol. (Unconfirmed rumour has it that this was a solo action of a junior political appointee.) Over time, and perhaps partly in response to the strong reaction from Europe, the Bush administration grew increasingly hostile to climate policy. Regardless of the position of the president, it is unlikely that the Kyoto Protocol would ever have been ratified by the Senate.

In 2001, in Marrakech, the finishing touches were put on the Kyoto Protocol. Emissions were defined (with an interpretation of carbon dioxide fluxes between the atmosphere and terrestrial biosphere that was very generous to Australia and Russia), no limits were set on the use of flexibility instruments, and no sanctions were imposed on violation of targets (or rather, countries in breach of their obligation will have to make up the gap, plus 30%, at some later time, on top of an unspecified future obligation).

After much toing and froing, the Kyoto Protocol came into force in 2005 after ratification by Russia (upon which the Kyoto Protocol imposes no obligations). Australia and Canada have been ambivalent about their commitments. Essentially, the Kyoto Protocol is a treaty between the European Union and Japan. Both are committed to climate policy in the absence of international treaties, so that the Kyoto Protocol is both narrow and shallow (in contrast to cartel theory which predicted that the Kyoto Protocol would be either narrow or shallow).

Since 2007, efforts have been undertaken to negotiate a successor to the Kyoto Protocol. The Kyoto Protocol did not expire in 2013, but its emission reduction targets became obsolete. A roadmap was agreed in Bali in 2007. Hopes were high for the 2009 meeting in Copenhagen, but the disappointment was greater. It was agreed 2010 in Cancun to keep talking, and again in 2011 in Durban. Another roadmap was agreed in 2012 in Doha.

In sum, the international negotiations on climate policy confirm that it is difficult to agree on the provision of a global public good. There is by now a vested bureaucratic interest to keep trying, but chances are that future negotiations will be as futile as past negotiations.

 FURTHER READING

Joseph Aldy and Robert Stavins' *Post-Kyoto International Climate Policy: Implementing Architectures for Agreement* (2010) and Scott Barrett's *Why Cooperate? The Incentive to Supply Global Public Goods* (2010) are excellent treatises on international climate agreements. Richard Benedick's *Ozone Diplomacy: New Directions in Safeguarding the Planet* (1998) gives a good introduction to the realpolitik of international environmental negotiations, and David Victor's *Global Warming Gridlock: Creating More Effective Strategies for Protecting the Planet* (2011) provides further insight into international climate policy.

IDEAS/RePEc has a curated bibliography on this topic at http://biblio.repec.org/entry/tdi.html.

 EXERCISES

11.1 Free-riding, as discussed in Section 11.3, is evaluated with respect to the grand coalition. Evaluate the difference in costs and benefits between full cooperation and no cooperation at all.

11.2 Graphically represent the two-player game of Section 11.4 as an Edgeworth Box.

11.3 For the two-player LQ game of Section 11.4, derive the costs and benefits of cooperation for both players. How does this change with α and β? Interpret the results.

11.4 Assume that the first Conference of the Parties (COP) to the United Nations Convention on Climate Change, in 1995 in Berlin, had a 50% chance of success. Assume that each COP is an independent try. What is the probability that COP19, in 2013 in Warsaw, will be a success?

11.5 Read and discuss:

- **A. Michaelowa and F. Jotzo (2006), 'Transaction costs, institutional rigidities and the size of the clean development mechanism', *Energy Policy*, **33**, 511–523.
- **M. Wara (2007), 'Is the global carbon market working?', *Nature*, **455**, 595–596.
- **M. Wara (2008), 'Measuring the Clean Development Mechanism's performance and potential', *UCLA Law Review*, **55** (6), 1759–1803.
- ***J.C. Murdoch and T. Sandler (1997), 'The voluntary provision of a pure public good: The case of reduced CFC emissions and the Montreal Protocol', *Journal of Public Economics*, **63**, 331–349.
- ***S. Barrett (2008), 'Climate treaties and the imperative of enforcement', *Oxford Review of Economic Policy*, **24** (2), 239–258.
- ****D. Osmani and R.S.J. Tol (2010), 'The case of two self-enforcing international agreements for environmental protection with asymmetric countries', *Computational Economics*, **36** (2), 93–119.
- ****A.M. Ulph and D.J. Maddison (1997), 'Uncertainty, learning and international environmental policy coordination', *Environmental and Resource Economics*, **9**, 451–466.
- ****S. Barrett (2006), 'The strategy of trade sanctions in international environmental agreements', *Resource and Energy Economics*, **19**, 345–361.

12

Adaptation policy

TWEET BOOK

- People adapt to climate change, reducing the negative impacts and increasing the positive impacts. #climateeconomics
- Adaptation reduces the need to mitigate climate change, and mitigation reduces the need to adapt. #climateeconomics
- Mitigation requires international cooperation. Adaptation does not. In fact, most adaptation is local and private. #climateeconomics
- International involvement in adaptation is obsolete except for the bureaucrats and consultants involved. #climateeconomics
- International adaptation policy by and large ignores the lessons of decades of development policy. #climateeconomics
- Most adaptation is private, e.g., clothing during a heat wave. Some adaptation is collective, e.g., siestas. #climateeconomics
- Some adaptation requires regulation, e.g., allowing people into air-conditioned shopping malls during heat waves. #climateeconomics
- Some adaptation involves the public sector, e.g., health care during heat waves, or agricultural extension services. #climateeconomics
- Little adaptation is public. Coastal protection and shared water resources are exceptions to the rule. #climateeconomics
- The public sector may hinder adaptation, e.g., agricultural subsidies that lock farmers into past behaviour. #climateeconomics
- Uncertainty is a major issue for long-lived water infrastructure. Adaptation demands more robustness and flexibility. #climateeconomics
- Elsewhere, systems change much faster than the climate, and adaptation will be one of many changes. #climateeconomics

12.1 Adaptation versus mitigation**

Adaptation includes any action to make the negative impacts of climate change less bad, positive impacts better, or even turn negative impacts into positive ones. Adaptation occurs on all levels of decision making, from individuals to the United Nations. Adaptation occurs on all time-scales from seconds to millennia. Some adaptation is in response to past climate change. Other adaptation is in anticipation of future climate change.

Adaptation aims to make climate change less bad. Therefore, adaptation reduces the need for emission abatement. Vice versa, emission reduction leads to less climate change. Therefore, mitigation reduces the need for adaptation. Thus adaptation and mitigation are policy substitutes. The more you do of the one, the less you do of the other.

Note that there are people who confuse substitutes versus complements with corner solutions versus interior solutions. Because climate change cannot be fully avoided, the policy mix includes both adaptation and mitigation. That does not make adaptation and mitigation complements, though.

Adaptation was low on the policy agenda for a long time. This was primarily because policy makers were focused on greenhouse gas emission reduction, and thought that discussing adaptation would be tantamount to admitting defeat. This has now changed. It is now widely accepted in policy circles that climate change cannot be fully avoided. (There was never any doubt about this in academic circles.) Whereas effective mitigation depends on the actions of other countries, effective adaptation rarely does. As the difficulties with reaching an international agreement on emission targets became increasingly apparent, attention has shifted to adaptation.

Many governments have formulated national adaptation plans. Indeed, they are obliged to do so under the UNFCCC. There is financial and technical support for poor countries to help develop their national adaptation plans. And there is a nascent multilateral adaptation fund to co-finance adaptation in poor countries, on top of the adaptation financed through the system of development banks.

This has been a bonanza for consultants. The next section discusses whether adaptation to climate change is a public good that merits government intervention.

12.2 The government's role in adaptation**

The impacts of climate change are many and diverse. As adaptation is about altering those impacts, adaptation is diverse too. Instead of talking about generalities, a few examples are discussed below.

Climate change will make heat waves more common. During hot weather, you should wear light clothes, not exert yourself, keep out of the sun, and drink lots (but little alcohol). This is adaptation, but private. People know this, and have the appropriate incentives to take these measures. There is no need for a government plan that will tell you not to put on your winter coat when it is hot outside. (Parents should tell young children, though.)

There are collective elements to this. It makes sense to have a siesta during the hottest hours of the day, and to have the main meal late in the evening. These things are more easily arranged when everyone in the neighbourhood does the same. There is no need, though, for the government to tell you when you should eat your dinner.

The government may run awareness campaigns to help unsuspecting populations cope with extreme heat (although such campaigns are not always effective), but there is no need for government intervention.

There are public elements, however. Over 40,000 people died during the 2003 heat wave in France. Many medical professionals were on holiday then. There was no procedure to call them back. The people who could overrule procedure were on holiday too. Shopping malls are air-conditioned, but security personnel removed people who came to seek relief from the heat rather than shop. In crime-infested neighbourhoods, people are afraid to open their windows, but there was no extra police on the street. (Note that violent and sexual crime is more common during hot weather.) Poor people could not afford the electricity to power theirs fans, but electricity bills were not waived during the heat wave. Clearly, the government has a role to play, as a facilitator of private adaptation, as a regulator, and as a service provider.

Agriculture is our second example. Climate change would affect crop yields. Farmers could respond in a number of ways. Planting and harvest dates could be changed, as could the application of pesticide, fertilizer and irrigation. Different varieties or different crops could be planted, or farmers could seek alternative livelihoods. Seed companies and extension services could support farmers with advice. It is in the farmer's own interest to adapt as the alternative is a drop in income.

As with health and heat, the role of the government is limited. Extension services are often state-owned and -run (even though no public good is provided) and so is large scale irrigation. In other ways, the government hinders adaptation. Import tariffs distort international trade and discourage specialization in what is comparatively advantageous. Subsidies similarly distort the market, rewarding particular activities at the expense of others and shielding farmers from market signals. In the European Union, subsidies are particularly generous in disadvantaged areas. That is, farmers are encouraged to grow the wrong thing in the wrong place. Here, withdrawal is the best the government can do for adaptation.

Sea level rise is the third example. There are private elements to adaptation. Tourists do not need to be told that they should not go sunbathing on a beach that has been eroded by sea level rise. Otherwise, coastal protection is a public good. Protection of a lot would be ineffective or exceedingly expensive unless it is coordinated with the projection of adjacent lots. Lots further inland benefit from the protection of the sea front, and should therefore contribute to the cost of coastal protection. Information asymmetries justify building codes (to help protect against wind and water) and land zoning. The government should take the lead in adaptation.

12.3 Adaptation and development**

Chapter 6 argued that the poorer countries tend to be more vulnerable to climate change. The reasons include the structure of the economy and a lack of adaptive capacity. Economic development is therefore a key component of adaptation policy.

As indicated above, rich countries and multilateral organizations are now funding adaptation in poor countries, using funds that would otherwise have been used as development aid. Mitigation has crowded out official development aid for longer. See Figure 12.1. There are two problems with this. First, adaptation is mostly a private good. External funding therefore crowds out internal funding. Outside money for adaptation really is an income transfer – but as the money is taken from development aid, the effect is zero.

Second, generic development aid is crowded out by specific development aid for adaptation to climate change. This may not be the first priority for development, and the adaptation money would be misallocated. The money for adaptation is partly controlled by people who understand climate change and its impacts, but who do not necessarily understand development.

International funding for adaptation is earmarked for projects that are obviously adaptation, as the allocation of money needs to be approved by bureaucrats. Building dikes and digging irrigation canals are obvious forms of adaptation. It resembles, however, the paradigm that dominated development economics in the 1950s and 1960s. Then, underdevelopment was

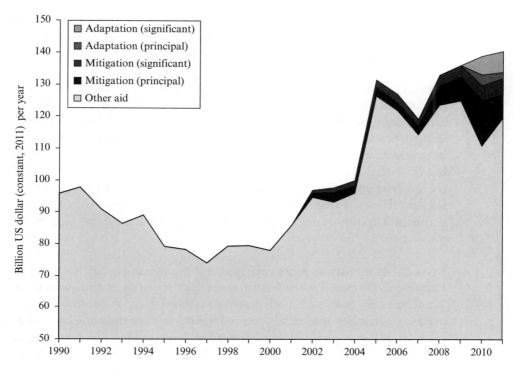

Figure 12.1 Total official development aid and aid for which adaptation and mitigation are the principal aim and a significant aim

deemed to be identical to undercapitalization, and development aid focused on infrastructure projects. This view is now known to be wrong. Providing information about future climate is another obvious form of adaptation. Western organizations operate satellites, databases and models to help African farmers make better decisions on crop choice and crop management. One could argue, however, that secure titles to land would do more to improve agriculture in Africa that better weather forecasts could.

12.4 How to adapt**

A number of examples of how to adapt to climate change are given above. In most cases, adaptation to climate change is like adaptation to any change. In many systems, climate changes more slowly than other drivers, so adaptation to climate change would not post any particular challenge.

There are two exceptions. Long-lived investments will have to withstand a wider range of climates. Future precipitation is particularly uncertain, with models disagreeing about the sign of change in many parts of the world. Investment in long-lived water infrastructure thus has to be prepared for a future in which anything can happen.

There are two ways of doing so: Make the investment more robust, or make the investment more flexible. Extra robustness entails that infrastructure can function under a wider range of weather conditions. This is relatively straightforward (but costly) if the sign of change is known. Sea walls, for instance, should be raised higher. The sign of future sea level rise is known, but the extent is uncertain. It would make sense to prepare for a high rate of sea level rise and raise the sea wall by more than is probably needed. This is because there is a large fixed cost in sea wall reinforcement (e.g., planning permission, transport disruption) but a relative low variable cost (e.g., materials, labour).

Extra flexibility may be more appropriate if the sign of change is unknown. Extra flexibility entails that infrastructure can be scaled up or down as needed. This comes at a cost too, as both design and materials are more advanced. For instance, a number of small reservoirs are more flexible than one big one, as the total storage capacity can be increased or decreased by commissioning or decommissioning one of the reservoirs. Moveable dams and inflatable barriers can be used to stop water only when needed. Retention areas can be used to temporarily store extra water. Infrastructure designed with such features command a premium in the face of an uncertain future.

 FURTHER READING

Every six years, Working Group II of the Intergovernmental Panel on Climate Change publishes a major assessment of the impacts of and adaptation to climate change. The information is layered, with a Summary for Policy Makers with high-level information, Technical Summaries with more detail, and multiple chapters with a lot of detail and references to the underlying literature. These reports can be found at: http://www.ipcc.ch/.

IDEAS/RePEc has a curated bibliography on this topic at http://biblio.repec.org/entry/tdj. html.

 EXERCISES

12.1 Consider an economic agent who seeks to maximize her consumption

$$\max_{A,R} Y - \alpha I(1 - A)(1 - R) - 0.5\,\beta A^2 - 0.5\gamma R^2 \qquad (12.1)$$

where Y is income, I monetized impact, A is adaptation and R is emission reduction; α, β and γ are parameters. What is the optimal level of adaptation? What is the optimal level of emission reduction? If emission reduction were zero, how would that affect optimal adaptation? If adaptation were zero, how would that affect optimal emission reduction?

12.2 Suppose that the agent of exercise 12.1 receives a donation D to finance adaptation. How does that affect her decisions?

12.3 Read and discuss:

- **S. Fankhauser, J.B. Smith and R.S.J. Tol (1999), 'Weathering climate change: Some simple rules to guide adaptation decisions', *Ecological Economics*, **30** (1), 67–78.
- **K.A. Miller, S.L. Rhodes and L.J. MacDonnell (1997), 'Water allocation in a changing climate: Institutions and adaptation', *Climatic Change*, **35**, 157–177.
- ***E.T. Mansur, R.O. Mendelsohn and W. Morrison (2008), 'Climate change adaptation: A study of fuel choice and consumption in the US energy sector', *Journal of Environmental Economics and Management*, **55**, 175–193.
- ***K.C. de Bruin, R.B. Dellink and R.S.J. Tol (2009), 'AD-DICE: An implementation of adaptation in the DICE model', *Climatic Change*, **95** (1–2), 63–81.
- ****R. Hasson, A. Lofgren and M. Visser (2010), 'Climate change in a public goods game: Investments decision in mitigation versus adaptation', *Ecological Economics*, **70** (2), 331–338.
- ****R.O. Mendelsohn (2000), 'Efficient adaptation to climate change', *Climatic Change*, **45** (3–4), 583–600.
- ****D.L. Kelly, C.D. Kolstad and G.T. Mitchell (2005), 'Adjustment costs from environmental change', *Journal of Environmental Economics and Management*, **50**, 468–495.

13

Building an integrated assessment model

13.1 Introduction

In this chapter, we will construct an integrated assessment model, that is, a model that combines the natural science and economic aspects of the climate problem to shed light on policy choices. The model will be built in Excel, an environment all have access to and most are familiar with. Model construction is done in 12 steps, corresponding to Chapters 2–13. The data needed to do this are available at https://sites.google.com/site/climateconomics/mliam.

On the same site, there is a version of the same exercise using Matlab. Detailed instructions are not given, because the Excel instructions below suffice for someone who can programme in Matlab.

13.2 Carbon cycle and climate

We start with two components: a carbon cycle model and a climate dynamics model.

The input (or *forcing*) to the carbon cycle model are annual emissions of CO_2, measured in Mt C (megatonne carbon = 1 million metric tonne carbon). The output of the carbon cycle model is atmospheric concentrations of CO_2, measured in ppm (parts per million).

The input (or *forcing*) to the climate model is the atmospheric concentration of carbon dioxide, again measured in ppm. The output of the climate dynamics model is the yearly average temperature increase over pre-industrial temperatures in °C.

The two components are coupled via the atmospheric concentration of CO_2, i.e., the output of the carbon cycle model is an input to the climate dynamics model.

13.2.1 Carbon cycle module

The carbon cycle model is a simple five-box model. The five boxes do not correspond to anything in the physical world; they are a mathematical abstraction that as a whole mimics the results from much more complicated models. In this model, all atmospheric CO_2 concentrations live in one of five boxes. If you want to compute the total atmospheric CO_2 concentration at a point in time, you simply add the amount of CO_2 in each of the five boxes. Over time, CO_2 *disappears* from each of these boxes, at different rates for each individual box. New anthropogenic CO_2 emissions are added each year into the atmosphere. In the five-box model these yearly influxes of new CO_2 into the atmosphere are distributed by fixed shares into the five boxes: 13% go into the first box, 20% into the second, 32% into the third, 25% into the fourth and the remaining 10% into the fifth box.

There are consequently five *variables* that represent the five boxes and each of these variables takes on a different value in each year. The equation that is used to compute the amount of CO_2 in box i (which takes values from 1 to 5) at time t (which takes on value from 1750 to 2008) is:

$$Box_{i,t} = \alpha_i Box_{i,t-1} + \gamma_i \beta E_t \qquad (13.1)$$

$Box_{i,t}$ is the amount of CO_2 in box i at time t, measured in ppm. α_i is the share of CO_2 in box i that stays in the atmosphere until the next time period (so $1 - \alpha_i$ is the share of CO_2 that disappears each year from box i). γ_i is the share of emissions that goes into box i. β is a unit conversion factor: CO_2 emissions in our model are measured in Mt C, but atmospheric CO_2 concentrations are measured in ppm; β converts from the unit Mt C to CO_2 ppm. E_t are world total emissions of CO_2 in year, t measured in Mt C.

The values for the emissions E_t are provided to you as an Excel file. You should use the following values for α_i and γ_i:

$\gamma_1 = 0.13$	$\gamma_2 = 0.20$	$\gamma_3 = 0.32$	$\gamma_4 = 0.25$	$\gamma_5 = 0.10$
$\alpha_1 = 1$	$\alpha_2 = \exp\left(-\dfrac{1}{363}\right)$	$\alpha_3 = \exp\left(-\dfrac{1}{74}\right)$	$\alpha_4 = \exp\left(-\dfrac{1}{17}\right)$	$\alpha_5 = \exp\left(-\dfrac{1}{2}\right)$

$Box_{i,1750} = 0$ except $Box_{1,1750} = 275$; $\beta = 0.00047$

The final step in the carbon cycle model is to compute atmospheric CO_2 concentrations at each point in time:

$$C_t = \sum_{i=1}^{5} Box_{i,t} = Box_{1,t} + Box_{2,t} + Box_{3,t} + Box_{4,t} + Box_{5,t} \quad (13.2)$$

C_t is the atmospheric concentration of CO_2 at time t, it is simply the sum of the five boxes at that time.

13.2.2 Climate module

The climate model has two parts: the first part computes the extra energy in the atmosphere and the long term temperature effect. The second part computes the yearly temperature increase over time.

The amount of extra energy caused by rising CO_2 concentrations is called the *radiative forcing* and is measured in W/m^2. The equation to compute it is

$$F_t = 5.35 \ln \frac{C_t}{C_{pre}} \quad (13.3)$$

F_t is the radiative forcing at time t caused by CO_2 in W/m^2. C_t is the atmospheric CO_2 concentration at point t in ppm, as computed by the previous component. C_{pre} is the pre-industrial level of atmospheric CO_2 concentrations, and you should use 275 ppm for this. $\ln(x)$ is the natural logarithm.

The next step in the model is to global mean surface air temperature. The equation for that is

$$T_t^A = T_{t-1}^A + \lambda_1 (\lambda_2 F_t - T_{t-1}^A) + \lambda_3 (T_{t-1}^O - T_{t-1}^A) \quad (13.4)$$

Here T_t^A is the increase in global average surface temperature; $\lambda_2 = 1.15$ is a parameter; $5.35 \lambda_2 \ln(2) = 4.26$ is the *climate sensitivity*, the equilibrium warming due to a doubling of the atmospheric concentration of carbon dioxide; $\lambda_1 = 0.0256$ is a parameter that determines how fast the atmosphere responds to a deviation between the actual and the equilibrium temperature; $\lambda_3 = 0.00738$ is a parameter that determines how fast the atmosphere responds to a deviation between the temperature of the atmosphere and ocean; and T_t^O is the temperature of the ocean, which follows:

$$T_t^O = T_{t-1}^O + \lambda_4 (T_{t-1}^A - T_{t-1}^O) \quad (13.5)$$

where $\lambda_4 = 0.00568$ is a parameter that determines how fast the ocean responds to a deviation between the temperature of the atmosphere and ocean.

13.2.3 Exercises

13.2.1. What happens to projected temperature if CO_2 emissions were held constant at 2008 levels in the model? What happens if CO_2 emissions grow by 2% per year? Hint: Introduce the annual growth rate of CO_2 emissions as a parameter. Store the results as values.

13.2.2. By how much would we need to reduce emissions from 2008 to keep global warming below 2°C in the year 2300? And over the next 300 years?

13.2.3. If emissions grow by 2% per year until the year 2030, and are then reduced by a fixed percent each year, how much would they have to be reduced in percent in each year to keep global warming below 2°C in the year 2300? And over the next 300 years?

13.3 Scenarios

You will add two components to your integrated assessment model: one that computes anthropogenic CO_2 emissions over time and an economic growth component that forecasts economic growth, both for three regions: rich, middle-income, and poor.

The output of the emissions component replaces the arbitrary emissions scenario for the carbon cycle component of Section 13.2. That is, instead of using emissions provided in a data file, you will compute emissions and couple the carbon cycle component to the emissions component. The growth component computes output, or GDP, and that will be an input into the emissions component.

13.3.1 Emissions model

The emissions component starts out with the Kaya identity:

$$M_{r,t}^* = P_{r,t} \underbrace{\left(\frac{Y_{r,t}}{P_{r,t}}\right)}_{\substack{\text{Percapita} \\ \text{GDP}}} \underbrace{\left(\frac{E_{r,t}}{Y_{r,t}}\right)}_{\substack{\text{Energy} \\ \text{intensity} \\ \text{of output}}} \underbrace{\left(\frac{M_{r,t}^*}{E_{r,t}}\right)}_{\substack{\text{Emission} \\ \text{intensity} \\ \text{of energy}}} \tag{13.6}$$

$M_{r,t}^*$ are industrial CO_2 emissions in Mt C in region r at time t, $P_{r,t}$ is population in region r at time t, $Y_{r,t}$ is output (or GDP) in region r at time t and $E_{r,t}$ is primary energy use in region r at time t.

The data file has population numbers for 1960–2010. If we assume, from 2011 onwards, that the population growth rate is 0.95 times the population growth rate in the previous year, then the world population stabilizes around 8.5 billion people.

The data file also has output for 1960–2010. Compute output per capita and its growth rate. For now, just assume that output per capita continues to grow at its 2010 rate. Compute total output for 2011–2300.

The data file further has primary energy use for 1960–2010 or a slightly shorter period, depending on the region. Compute the energy intensity and its growth rate. Assume that energy intensity continues to fall at the average rate over the entire period for which there are observations. Compute total primary energy use for 2011–2300.

Finally, the data file has carbon dioxide emissions for 1960–2008. Compute the carbon intensity and its growth rate. Assume that carbon intensity continues to fall at the average rate over the entire period for which there are observations. Compute total carbon dioxide emissions for 2009–2300.

At this point we have computed emissions from industrial activities, assuming no specific climate policy is implemented. The economic part of the scenarios needs to be improved, though.

13.3.2 Growth model

You will build a simple growth model – the Solow model – that will replace output in the emissions component above.

Output in a specific year is computed by a production function that depends on three things: the amount of capital (i.e., factories, machines etc.), the amount of labour (in our case equal to the population size) and a technology index, also called total factor productivity, i.e., a measure of how efficient we are in using the inputs capital and labour to produce things. The production function used is called a Cobb–Douglas production function. It has the following form:

$$Y^G_{r,t} = A_{r,t} K^{\alpha}_{r,t} P^{1-\alpha}_{r,t} \tag{13.7}$$

$Y^G_{r,t}$ is gross output in trillion dollars (we discuss why it is called "gross" output in Section 13.4) in region r at time t. $A_{r,t}$ is the total factor productivity in region r at time t. $K_{r,t}$ is the amount of capital available for production in region r at time t. $P_{r,t}$ is population available for production, measured in trillion dollars. Parameter $\alpha = 0.2$ is called the capital share.

The capital stock is modelled in a similar way to our modelling of the concentration of CO_2 in the atmosphere: we assume that there is an inflow of new capital (i.e., new factories and machines are built) and that some capital breaks over time, so there is an outflow of capital. The equation of motion for the capital stock is given as:

$$K_{r,t} = (1 - \delta)K_{r,t-1} + I_{r,t-1} \tag{13.8}$$

$\delta = 0.1$ is the depreciation rate of capital. $I_{r,t}$ is investment, i.e., a measure of how much new capital is built every year.

The amount of new investment into capital for year t should be modelled as:

$$I_{r,t} = sY_{r,t} \tag{13.9}$$

$s = 0.2$ is called the *savings' rate*.

As with all equations of motion, you cannot use Equation (13.8) to compute the level of the capital stock for the initial period. Instead, we initialise it at its steady state:

$$K = (1 - \delta)K + sAK^{\alpha}P^{1-\alpha} \Leftrightarrow \delta K = sAK^{\alpha}P^{1-\alpha} \Leftrightarrow \tag{13.10}$$

$$\delta K^{1-\alpha} = sAP^{1-\alpha} \Leftrightarrow K = \left(\frac{sA}{\delta}\right)^{1/1-\alpha} P$$

The initial level of total factor productivity is found by calibration. Start with a value of 1 and change it until modelled output equals observed output in 1960. Alternatively, use the Solver in Excel (see Section 13.9) to find the value that sets the difference to zero.

Let total factor productivity grow at a constant rate between 1960 and 2010. Start with a value of 2%. Use the Solver to find the growth rate that ensures that the modelled output in 2010 equals the observed output.

For 2011–2300, let the growth rate of total factor productivity equal 0.99 times the total factor productivity growth in the previous period.

13.3.3 Coupling

At this point we can couple the growth component with the rest of the model. First, replace the arbitrary growth rate of per capita income. Instead, let total output grow as per the Solow model. Per capita income then follows.

Second, let the emissions that drive the carbon cycle model, grow at the same rate as the modelled emissions.

13.3.4 Exercise

13.3.1. Decompose the main drivers of climate change in the scenario used in the model and rank their relative contribution to future climate change.

First, add one period to the model. The equations will be safely stored in the year 2301. Change the equations for the years 2011–2300. Save the results. Change the equations back by copying the equations for 2301 to 2011–2300.

Copy and paste/values of the global mean temperature in the exercises sheet. Then assume that population is frozen at its 2010 level. Store the results in the exercises sheet. Restore population to its scenario values. Repeat this exercise with per capita income frozen, energy intensity frozen, and carbon intensity frozen.

Briefly describe the effect each of these has on average temperatures, and rank them in terms of the size of their effect.

13.4 Abatement

The emissions module of Section 13.3 lacks an option for greenhouse gas emission reduction. In order to model this, we will have to introduce a so-called choice variable: the emission control rate. This is a new type of variable. Unlike a scenario variable that is an external driver, or a state variable, which is computed by some equation, a choice variable is something for which we use our model to find an "appropriate" value. The emission control rate is such a choice variable. We want to use our model to compute the amount we should reduce CO_2 emissions per year in order to reach a given objective (and most often that objective will be some balancing act of costs and benefits of reducing emissions). For now, just introduce a new row in your Excel spreadsheet for this choice variable, it will have a value for each year and you can initially set the control rate to 0% (i.e., no climate policy). We will designate the emission control rate by $R_{r,t}$.

The final equation for emissions therefore is

$$M_{r,t} = (1 - R_{r,t})M^*_{r,t} \tag{13.11}$$

We assume throughout our model that carbon policy is costly. So any choice of $R_{r,t}$ larger than 0% will impose a burden on the economy. We compute the relative size of this burden, also called relative abatement cost, by a simple function:

$$B_{r,t} = \beta R_{r,t}^2 \tag{13.12}$$

$B_{r,t}$ is the relative cost of climate policy at time t, $\beta = 0.1$ is a parameter.

In addition to $B_{r,t}$, you should add another variable to your Excel sheet that computes the cost of climate policy in trillion dollars.

We called output so far gross output because it does not account for the cost of climate policy. The equation for net output includes abatement costs and is given as

$$Y_{r,t} = (1 - B_{r,t}) Y_{r,t}^G \qquad (13.13)$$

So this equation picks up the effect of the control variable from the emissions component. The absolute cost of emission reduction is then given by $B_{r,t} Y_{r,t}^G$.

13.4.1 Exercises

13.4.1. Vary $R_{r,t}$ between 0%, 5% and 10% for the period 2015–2399. What happens to emissions, concentrations, and temperature? What are the costs of these policies?

13.4.2. Compute the marginal abatement costs. (Hint: Substitute Equations (13.11) and (13.12) into Equation (13.13) to compute total emission reduction costs as a function of relative emission reduction. Rescale the β parameter so that the units match. Then take the first partial derivative of the total costs to emissions.) What are the regional emission reduction rates if the marginal abatement costs are equal for the three regions and global emission reduction is 5% or 10%? Do this for 2015–2020 only, and keep uniform emission reduction elsewhere. Impose these emission cuts. What is the difference in costs with the previous exercise?

13.4.3. In the previous exercise, you computed the marginal abatement costs in each period for a given emission reduction. Keep the marginal abatement costs as is in the first period, and compute the emission reduction in later periods such that the marginal abatement cost rises with the rate of discount. What are implications for emissions, concentrations and temperature? Change the emission reduction in period 1 such that the temperature in 2100 is the same as in the first exercise. What is the difference in costs?

13.5 Tradable permits

You will add a component to the model to simulate an international market in emission permits.

As above, each region faces an abatement cost function that is quadratic in relative emission reduction – see Equation (13.12) – and linear in emission permit purchases and sales:

$$B_{r,t} = \beta R_{r,t}^2 Y_{r,t} + \pi P_{r,t} \qquad (13.14)$$

Each region has an emissions target $T_{i'}$ and minimizes abatement costs by choosing the optimal amount of in-house emission reduction and permit sales. Assume that there is an international market in emission permits. Assume that all companies are price-takers $-\frac{\partial \pi}{\partial R} = 0$ – and derive the market equilibrium. Implement this in the Excel model.

13.5.1 Exercises

13.5.1. Assume that μ_t is 5% or 10%. This sets the initial allocation of permits. Now impose trade. What happens to total and marginal abatement costs? How does this compare to the solution with a tax?

13.5.2. Assume that the poorest region has no emission reduction obligations, and halve the obligations of the middle-income regions. Increase the abatement obligations of the richest region such that global emissions are as above. (Hint: The emissions target for the rich region follows from the global target and the targets of the other two regions.) What are the marginal and total costs without emission permit trade? And with? How does this compare with the previous exercise?

13.6 Impacts of climate change

We now add a component for the impacts of climate change to your model. You also need to modify the growth model to pick up the estimate of climate impacts. With that component, you have a model that provides estimates of everything needed to do a cost–benefit analysis of climate change policy.

13.6.1 Impact model

The impact model is simple. We assume that the harm done from rising temperatures in a given year in percent of output is

$$D_t = \psi_1 T_t + \psi_2 T_t^2 + \psi_6 T_t^6 \tag{13.15}$$

where D_t is impact in percent of output in year t, T_t is global average temperature in °C above pre-industrial levels at time t and ψ_1, ψ_2 and ψ_6 are parameters that you should set to:

Region	Model 1			Model 2		
	Rich	Mid	Poor	Rich	Mid	Poor
ψ_1	5.88	3.57	1.96	0	0	0
ψ_2	−2.31	−1.70	−1.26	0.5563	0.2561	0.0655
ψ_6	0	0	0	−0.0113	−0.0106	−0.0101

You should pick up the temperature from the climate dynamics component you have built previously.

13.6.2 Growth model

To close the loop, you should modify the equation for net output in the growth model to not only subtract the costs of abatement from gross output, but also the damages from climate change that you just added to the model. You have to figure out the precise new equation for net output yourself. Have a look at how abatement costs were subtracted from gross output and try to do the same for the impacts of climate change.

The last addition to our model is two variables: consumption (in trillion dollars) and per capita consumption (in dollars per person). The equation for consumption is straightforward: whatever is left of gross output once the costs of abatement, the damages from climate change and investment in the capital stock are subtracted can be consumed and thus equals consumption. To compute per capita consumption you divide consumption by population. Be careful with the units, though!

And last, but not least, make sure that your emissions component is coupled to the economic growth component (by setting that switch to the correct value) before you answer the policy questions.

13.6.3 Exercises

13.6.1. For Model 1, find out in which year the net change in per capita consumption from four different policies turns beneficial. The policies you should analyze are characterized by a constant reduction in emissions, i.e., the same percentage reduction in each year. The four policies you should analyze are a 5%, 10%, 15% and 20% emission reduction. For each policy, compute how the per capita consumption in each year changes compared with the no-policy scenario. For each policy, find the first year in which per capita consumption is higher with policy compared with the no-policy case.

13.6.2. Create a graph that plots the change in per capita consumption for all four policies, one that plots abatement costs as percent of output and one that plots damages as percent of output. The horizontal axis should be years in all three figures. The vertical axes should be percentage change in per capita consumption for the first figure, and percent of output for the second and third figure. Each policy should be one line in each figure.

13.6.3. Repeat the first exercise but now with ψ_1, ψ_2 and ψ_6 according to Model 2.

13.7 Social cost of carbon

We can now use the model to compute the social cost of carbon (SCC). Recall that the social cost of carbon is the net present value of the impact caused by an extra emission of 1 tonne of carbon today.

The general strategy for computing the SCC is as follows: you run your model twice. The first run (base run) is identical to the model set-up as you have been using it. In the second run there should be an additional emission of 1 tonne of carbon into the atmosphere in the year 2015. This second run (the marginal run) will therefore have slightly more warming, and that will cause slightly larger damages from climate change.

You then compute the difference in damages between the base and marginal run for each year in dollars. We call this "marginal damages", i.e., this is the time series of additional damages caused by one additional tonne of carbon emitted today.

The next step is to compute the net present value of marginal damages. This is a simple step: you just multiply the marginal damages in each year with the discount factor for that year. This gives you a new time series of the net present value of marginal damages. You will do this for different discounting schemes that are described in more detail below. So in practice you will have a separate time series of discount factors for each of the schemes, and then a separate time series of net present values of marginal damages for each scheme.

The final step is to add up the net present value estimates of marginal damages over time for each of the discounting schemes. This will give you one number for each discounting scheme. That number is called the social cost of carbon.

13.7.1 Some practical advice

You should start by adding columns for the perturbed emissions, concentrations, temperatures and impacts. You should simply add to the emissions, i.e., don't try to modify your Kaya identity. You should disregard the differential feedback on economic growth. Just keep growth as it is. This makes programming easier, and it avoids some potential problems in welfare analysis.

13.7.2 Discount factors

In total you will use six different discount factor time series, which will give you six different estimates of the social cost of carbon. They are split into two groups: the first three are based on a constant discount rate, the last three are based on the Ramsey rule.

For the constant discount rate the equation for the discount factor is

$$DF^c_t = \frac{1}{(1 + r^c)^t} \tag{13.16}$$

Here DF^c_t is the discount factor for time step t. Careful, t here is not year, but the time step counted from the start of the model, so $t = 0$ corresponds to the year 2015, $t = 1$ to 2016 and so on. r^c is the discount rate. You should use the constant discount rate approach for three different discount rates: 2%, 3% and 5%. These three rates constitute the first three discount schemes.

The Ramsey discount rate is a bit more complicated. Because the discount rate for each year depends on that year's per capita consumption growth rate we first need to compute the Ramsey rate for each year:

$$r^r_t = \rho + \eta g_t \tag{13.17}$$

r^r_t is the Ramsey discount rate for year t, ρ is the pure rate of time preference (prtp), η is the rate of risk aversion, and g_t is per capita consumption growth from year $t = 1$ to year t (so $g_t = \frac{c_t - c_{t-1}}{c_{t-1}}$, with c_t being world average per capita consumption in year t). Due to the definition of the Ramsey discount rate you cannot compute it for the first year, so you should compute it for the second and all following years.

The equation for the Ramsey discount factor is

$$DF^r_t = \prod_{s=0}^{t} \frac{1}{1 + r^r_s} \tag{13.18}$$

This is tricky to put into Excel right away. Instead, you can use a recursive formulation that is mathematically equivalent:

$$DF^r_t = \frac{DF^r_{t-1}}{1 + r^r_t} \tag{13.19}$$

This gives you the discount factor for all but the first year. The discount factor in the first year is one by definition: $DF^r_0 = 1$.

You should compute three different discount schemes based on the Ramsey equation, one for each of three different values for the pure rate of time preference: 0.1%, 1% and 3%.

13.7.3 Exercises

13.7.1. Compute the social cost of carbon for the six alternative discount rates and the two alternative impact models.

13.7.2. Does the social cost of carbon change if we implement carbon policy? The original set-up computed the social cost of carbon assuming no climate policy. We can also ask the question: if we already reduce emissions by say 5% and 10% from 2015 onwards, how much additional damage is caused if we then reduce emissions by one extra tonne? Do this for the high discount rates and Model 1 only.

13.7.3. How does the SCC change for different values of the climate sensitivity? The IPCC states the equilibrium warming for a doubling of CO_2 concentrations is likely in the range of 2 to 4.5°C with a best estimate of 3°C. Note that in Section 13.2 we found a climate sensitivity of 4.26°C by calibration. The equation we use to compute the equilibrium warming is:

$$\Delta T_t = \lambda \times 5.35 \ln \underbrace{\frac{C_t}{C_{pre}}}_{\substack{\text{This is 2 for a} \\ \text{doubling of} \\ \text{concentrations}}} \tag{13.20}$$

We got the value for λ (the climate sensitivity parameter in our model) by plugging in the numbers from IPCC and solving for λ:

$$3° = \lambda \times 5.35 \ln 2 \Rightarrow \lambda = 0.8 \tag{13.21}$$

You should solve this equation to compute the climate sensitivity parameter that gives a warming of 2° and a warming of 4.5° for a doubling of CO_2 concentrations, and then run your model with these alternative climate sensitivity parameters. How does the SCC change for these alternative climate sensitivities? Do this for the high discount rates and Model 1 only.

13.8 Development

You now add one more twist to the model. Above, we assume that vulnerability to climate change is constant. At the same time, we assume that the poorer regions are more vulnerable. This is inconsistent. So, we introduce an income elasticity into Equation (13.15), as follows

$$D_t = (\psi_1 T_t + \psi_2 T_t^2 + \psi_6 T_t^6)\left(\frac{y_t}{y_0}\right)^\varepsilon \tag{13.22}$$

where $\varepsilon = -0.25$ and $y_0 = y_{2010}$.

13.8.1 Exercises

13.8.1. Vary ε between −0.50, −0.25, 0.00, and 0.25. What are the implications for the social cost of carbon?

13.8.2. Reconsider emission reduction of 5% in the richest region in 2015. Assume that the richest region does not reduce emissions but instead donates the money to the poorest region. What are the implications for emissions? What are the implications for the impact of climate change?

13.8.3. In the specification above, we assume that the impacts of climate change scale down output, or drive a wedge between gross and net output. This implies that the impacts of climate change reduce both consumption and investment. First, assume that all impacts fall on consumption. What are the implications for economic growth and emissions? Now let all impacts fall on investment. That is, keep consumption as it would have been without climate change, but reduce investment. What are the implications for economic growth and emissions?

13.9 Optimal climate policy

For this assignment, you compute the optimal climate policy trajectory over time. There are three steps that you need to finish in order to do so: (1) you need to add a welfare function to the model, (2) you need to set things up so that you can use a numerical optimization package that is part of Excel, and (3) you need to run this numerical optimization package to find the optimal policies for a variety of different assumptions.

13.9.1 Welfare component

The first step is to add a component that computes the overall social welfare for a given policy. This will be one number, and the policy that gives us the highest number for this metric is the one we will label "optimal".

The welfare function you should add as a component has the following equation:

$$SWF = \sum_{t=0}^{T} P_t \ln c_t \left(\frac{1}{1+\rho}\right)^t \qquad (13.23)$$

P_t is the population size at time t and c_t is per capita consumption at time t. Both are variables you are computing in other parts of the model already. ρ is the pure rate of time preference, and you should set the component up in a way that you can easily change its value. T is the time horizon of your model.

13.9.2 Preparing the model

The way our model is set up at this point allows us to set a different mitigation level (emission control rate) for each year of our analysis. This amounts to $3\times(2300-2015)=855$ decision variables, and the numerical optimization package we intend to use therefore needs to find the best value for each of these decision variables. This would not be a problem for state-of-the-art optimization packages, but it is a problem for Excel. The next step therefore is to transform our 855 decision variable problem into one that has only nine decision variables.

We will do this by creating nine new decision variables: for 2015–2019, 2020–2024, and 2025–2300, per region. Put these in a little matrix somewhere, and let emission reduction refer to this matrix.

The next step is to enable the numerical optimization package in Excel. The name of it is "Solver" and it is switched off by default. Here is how you can enable it in Excel 2010:

- Click on File->Options->Add-Ins
- Make sure "Excel Add-ins" is selected in the "Manage" field
- Click "Go. . ."
- Select "Solver Add-in"
- Click "Ok."

This adds an item "Solver" under Data->Analysis in the main Excel window, and you can start Solver clicking on that new item.

The next step is to tell Solver what cell it should try to maximize, and which cells it can modify in order to find the best combination of values for the decision variables. To do so, start Solver, and then select the cell with the value for your social welfare function for the "Set Objective" field. Make sure you have selected the option to maximize that cell (and not minimize it or find a specific value). Next, you need to select the range of your ten new decision variables for the field "By changing variable cells". At this point Solver knows that it should try different values for these ten cells, and try to find the combination of values that gives the highest value for the cell that you selected as the objective (in our case gives the highest social welfare).

Before we can run Solver, we need to tell it one more thing: what is the range of values that makes sense for our decision variables. In our case the decision variables are emission control rates that can take values from 0 to 1 (i.e., 0% to 100%). So we don't want Solver to try any values that are outside that range. We can configure that by setting up constraints in Solver. You will have to add two separate constraints, one that says the decision variable always has to be greater or equal to 0, and one that says it always has to be smaller or equal to 1. You can add a constraint by clicking the "Add" button. For the cell reference you then select the same cells that you already picked as the decision variables, then you need to select the correct condition (i.e., >= for the first and <= for the second constraint) and finally in the field constraint you simply add either 0 for the first and 1 for the second constraint. When you are done, you should have both constraints listed in the main window of Solver in the field "Subject to the Constraints".

Note that you may also try without constraints and see whether they are violated.

13.9.3 Exercise

13.9.1. Find the optimal policy for the two alternative specifications of the impact model. What are the implications for concentrations, temperature and per capita consumption?

13.10 Discounting and equity

You can now add equity weights to the model. Equity weights do not affect the inner workings of the model. They reinterpret the results. However, as the optimal carbon tax is affected, equity weights do affect the outcomes.

Equity weights are defined as

$$GWF = \sum_{r=1}^{R} \left(\frac{\bar{y}}{y_r} \right)^{\eta} SWF_r \qquad (13.24)$$

That is, equity weights are the ratio of global average per capita income to regional average per capita income, raised to the rate of risk aversion. Note that we first discount future impacts to today before we apply the equity weights.

13.10.1 Exercises

Hint: Use the model version of Section 13.8.

13.10.1. Compute the social cost of carbon with equity weights for $\eta = 0.0$, $\eta = 0.5$, $\eta = 1.0$, $\eta = 1.5$ and $\eta = 2.0$.

13.10.2. Change the pure rate of time preference from 3% per year to 4%, 2%, 1% and to 0.1% and re-compute the social cost of carbon (without equity weights).

13.10.3. Change the rate of risk aversion from η = 1.0 to η = 0.0, η = 0.5, η = 1.5 and η = 2.0. Only consider the implications for the discount rate. (Hint: Use the Ramsey rule.) Compute the social cost of carbon (without equity weights).

13.10.4. Now also consider the effect on equity weights. Compute the social cost of carbon.

13.11 Uncertainty

13.11.1 Exercise

13.11.1. Return to the model of Section 13.9. Reduce the number of control variables to 3×2: per region, abatement in 2015–2024 and abatement in 2025–2300. Optimize the emissions control rate for climate sensitivities of 3.0°C/2×CO$_2$, 1.5°C/2×CO$_2$, and 4.5°C/2×CO$_2$.

Modelling uncertainty is not that difficult but the two-dimensional representation of the model in Excel gets in the way. So far, we have worked in two dimensions: Different years were found in different rows, and different variables in different columns. We now need a third dimension: State of the world. So far, we had one state of the world. We assumed that variables and parameters were perfectly known and therefore could be represented by a single number. We now introduce three states of the world for one parameter: The climate sensitivity. Previously, this was 3.0°C/2×CO$_2$. Here, it has three alternative values: 3.0°C/2×CO$_2$ with a 75% probability, 1.5°C/2×CO$_2$ with a 10% probability and 4.5°C/2×CO$_2$ with a 15% probability.

This means that you need to split the column that contains the atmospheric temperature variable into three: one low temperature, one middle, and one high. This also means that you need to split every variable that depends on the temperature, directly or indirectly, into three. Because we have built an integrated assessment model in which everything depends on everything, the entire model needs to be split. For instance, emissions depend on economic output, and economic output depends on climate change.

So, it is best to create three separate sheets, each containing the same model, but one with a low climate sensitivity, one with a middle climate sensitivity, and one with a high climate sensitivity.

You now understand why we have only three states of the world, with three alternative values for one parameter. We could have had seven or seven hundred alternative values for the climate sensitivity, but then we would have needed seven (hundred) sheets. We could have had two uncertain parameters (in fact, every parameter in the model is uncertain), but then we would have needed 3×3=9 sheets (if both parameters can assume only three alternative values). Other programming environments than Excel have data structures that are more amenable to uncertainty analysis.

We have created three separate sheets with three alternative versions of the model. There are two components, however, that are common: Emission reduction and expected utility. Optimal emission reduction follows from maximising the expected value of the net present value of utility. Expected net present utility equals 0.10 times net present utility in the low model, plus 0.75 times net present utility in the middle model, plus 0.15 times net present utility in the high model. Compute expected net present utility either in a fourth sheet or in the middle sheet. Introduce a variable for the annual emissions control rate. Set the control rate in each of the three model variants equal to the common control rate.

13.11.2 Exercise

13.11.2. Optimize the emissions control rate under uncertainty. Compare the results with the emissions control rate under certainty of Exercise 13.11.1. Repeat the exercise with probabilities 0.15/0.65/0.20 and with 0.05/0.75/0.20.

Let us now assume that the truth about the climate sensitivity will be revealed in 2025. This will not happen, but this assumption teaches us something about the impact of learning on optimal emission reduction. We do not know what truth will be revealed, but we do know that it will be.

This changes the way we set up the optimization. First, copy the existing spreadsheet to a new one. From 2025 onwards, we have three separate optimizations. The emissions control in the low model is set by maximizing the net present value (in 2025) of utility in low model. Ditto for the middle and high models.

For the period 2015–2024, the emissions control rate is set by maximizing the expected net present value *over the entire period*. Thus decisions on the emission control rate in 2015–2024 depend on decisions about the control rate in 2025–2300. Vice versa, emission control in 2025–2300 depends on emission control in 2015–2024. Emissions and atmospheric concentration of carbon dioxide in 2025 obviously matter, but the temperature has inertia too.

One way to solve this is by iteration. Take the control rate without learning as the starting point. Optimize 2025–2300 assuming that 2015–2024 as is without learning. Optimize 2015–2024. Reoptimize 2025–2300. Reoptimize 2015–2024. And so on until nothing much changes.

Alternatively, because emission control in one state of the world does not directly affect emission control in another state of the world, we can just maximize a weighted sum of welfare in alternative states of the world. In the same optimization, we can include 2015–2024. There are now four control variables per region: 2015–2024 and low / mid / high for 2025–2300.

13.11.3 Exercise

13.11.3. Optimize the emissions control rate under uncertainty and learning. Compare the results with the emissions control rate under certainty, and under uncertainty.

13.12 Non-cooperative climate policy

Let us now return to the deterministic model of Exercise 13.11.1. There, we maximized the net present utility of the world as a whole, which was equal to the sum of the net present utility of the three regions. Each region had its own emission control rate, because the first order conditions have that the marginal abatements costs are equalized (rather than the control rate).

13.12.1 Exercises

13.12.1. Optimize the emissions control rate separately for each of the three regions by maximizing net present regional utility. Do this iteratively. In the first iteration, assume that the other regions do not reduce their emissions. In later iterations, assume that other regions reduce their emissions as in the previous iteration. Repeat until convergence. Compare the results of this non-cooperative solution with the cooperative solution of Section 13.11 (under certainty).

13.12.2. Compare the welfare levels of the three regions with and without cooperation. Can the winner of cooperation compensate the losers if welfare can be transferred? What if welfare is not transferable but money is? (Hint: A welfare change times the inverse of marginal utility is the willingness to pay or willingness to accept compensation in dollar terms.)

13.13 Adaptation policy

Equation (13.22) gives the economic impact of climate change. It makes implicit assumptions about adaptation. Let us make the assumptions explicit. We define gross impact G as the impact without adaptation A, and residual impact as $G(1+A)$. If G and A have the same sign, adaptation increases positive impacts and reduces negative impacts. Total impact is then residual impact minus adaptation costs. More specifically,

$$D_t = \varphi T_t (1 - A_t) - \chi_1 A_t^{\chi_2} \tag{13.25}$$

where A is adaptation effort. The first term on the right-hand side is now gross impact, or impact before adaptation. Impact now equals residual impact, or impact after adaptation, plus adaptation costs. Note that we need to define χ_1 such that adaptation costs are defined as a share of output.

The optimal level of adaptation follows from

$$\frac{\partial D_t}{\partial A_t} = -\varphi T_t - \chi_1 \chi_2 A_t^{\chi_2 - 1} = 0 \Rightarrow A_t = \left(\frac{-\varphi T_t}{\chi_1 \chi_2} \right)^{\frac{1}{\chi_2 - 1}} \tag{13.26}$$

Because gross impact can be positive as well as negative, χ_2 has to be a natural number. Because gross impact and adaptation need to have the same sign, χ_2 has to be an even number. So $\chi_2 = 2$. The other parameters are:

	rich	middle	poor
ϕ	−5.46	−5.05	−4.43
χ_1	59.5	36.4	32.0
ψ_1	−5.46	−5.05	−4.43
ψ_2	0.13	0.17	0.15

13.13.1 Exercises

13.13.1. Optimize the emissions control rate with explicit adaptation. Compare the results with the emissions control rate with implicit adaptation using a quadratic impact function $\psi_1 T + \psi_2 T^2$.

13.13.2. Assume that there is an international adaptation fund: Transfer $100 billion dollars from the rich region to the poor region. How does this affect adaptation decisions in the poor region?

14

How to solve the climate problem?

TWEET BOOK

- Putting more greenhouse gases in the atmosphere will change the climate but it is uncertain how and how much. #howtosolvetheclimateproblem
- Climate change has positive and negative impacts. Net effect is negative but small compared with economic growth. #howtosolvetheclimateproblem
- A gradually rising carbon tax reduces emissions at minimum cost. Cost would be small for reasonable target. #howtosolvetheclimateproblem
- Climate alarmism meets the religious demand for eternal doom, sinful emissions, and atonement. #howtosolvetheclimateproblem
- Climate policy allows politicians to promise the world, postpone major action, and blame Johnny Foreigner. #howtosolvetheclimateproblem
- Climate policy lets bureaucrats build new bureaucracies. It feeds fears of right-wing conspiracy theorists. #howtosolvetheclimateproblem
- Greenhouse gas emission reduction is a global public good. It is better if someone else does it. #howtosolvetheclimateproblem
- There is a clear and sustained public demand for climate policy, even if it means more expensive energy. #howtosolvetheclimateproblem
- Abatement is easier if in step with trade partners. UNFCCC data standards plus pledge and review are enough. #howtosolvetheclimateproblem
- Abatement is easier if it can be bought wherever it is cheapest. Kyoto Protocol allows for this. #howtosolvetheclimateproblem
- There is not enough conventional oil and gas to cause substantial climate change. Alternatives might. #howtosolvetheclimateproblem
- Climate policy should ride with, rather than against, the ongoing revolution in energy supply. #howtosolvetheclimateproblem

14.1 The problem

Greenhouse gases are transparent to visible light but not to infrared radiation. Energy from the sun thus easily enters the planet. Energy re-emitted by Earth finds it more difficult to leave: It is absorbed by greenhouse molecules in the atmosphere, and scattered in any direction – including back to the surface. This is the natural greenhouse effect. See Figure 1.3. Planet Earth is warmer than it would have been without greenhouse gases in the atmosphere. If greenhouse gas concentrations increase, then you would expect from first principles that the planet would become hotter.

The concentrations of three of the main greenhouse gases, carbon dioxide (CO_2), methane (CH_4) and nitrous oxide (N_2O), have increased steadily since the start of the Second Industrial Revolution (1750 say). The increase is dramatic if we consider that greenhouse gas concentrations had been more or less stable since the last Ice Age and the Agricultural Revolution. See Figure 1.1. The increase in concentrations is no surprise as greenhouse gas emissions are associated with fossil fuel combustion and deforestation (CO_2), with population growth and affluence (via meat and rice production and waste generation CH_4), and with artificial fertilizers (N_2O).

Over the course of the 20th century, a rise in temperature has been observed, as well as a decrease in snow cover, and a rise in sea level (due to thermal expansion as water warms). See Figure 1.2. The impact of the enhanced greenhouse effect on the climate does not follow from first principles alone. The climate is a complex system. Any initial change sets in motion a cascade of feedback effects, some positive and some negative. The most powerful feedbacks relate to water. A warmer atmosphere contains more water vapour, and water vapour is a powerful greenhouse gas. Cloud formation would be affected. Clouds can either cool – e.g., on a summer day – or heat – e.g., in a winter night. Ice is white and reflects sunlight. Water is dark and absorbs lights. Climate is also affected by a range of other things, some natural – variations in solar radiation, volcanoes, ocean dynamics – and some human – aerosols, land use change, nutrients.

State-of-the-art climate models include these feedbacks and many more. These models project that the warming observed in the 20th century will continue during the 21st century and beyond. Models differ on the detail, though, and the range of future projections is enlarged because emission projections are highly uncertain too. See Figure 1.7. Besides warming, climate change would also entail changes in wind and precipitation patterns.

14.2 Costs and benefits of climate policy

Some people argue that climate change is bad, as all change is for the worse. This is an odd position. Universal education for girls would be a radical departure from the past, but is generally welcomed. Research has shown that climate change would bring both positive and negative impacts. Positive impacts include a reduced demand for energy for winter heating, fewer cold-related deaths, and CO_2 fertilization which makes crops grow faster and reduce their demand for water. Negative impacts include sea level rise, the spread of tropical diseases, and increases in storm intensity, droughts, and floods.

Adding up all these impacts after having expressed them in welfare equivalents, the impact of initial climate change is probably slightly positive. This is irrelevant for policy, because initial climate change cannot be avoided. More pronounced climate change would have net negative effects, and these impacts would accelerate with further warming. Even so, the impacts would be moderate: The welfare impact of a century of climate change is comparable with the welfare impact of a year of economic growth. Uncertainties are large, though, but even the most pessimistic estimates show that a century of climate change is comparable with a decade of growth. See Figure 6.2.

Greenhouse gas emissions can be reduced in a number of ways. More efficient energy use and a switch to alternative energy sources are the two main options. This is best stimulated by a carbon tax. Incentive-based policy instruments are better suited for reducing emissions from diffuse and heterogeneous sources than rule-based instruments. Taxes are more appropriate for stock pollutants than tradable permits. A carbon tax is therefore the cheapest way to reduce greenhouse gas emissions.

Net present abatement costs are lowest if all emissions from all sectors and all countries are taxed equally and if the carbon tax rises with the interest rate. Higher carbon taxes would lead to deeper emission cuts. Only a modest carbon tax is needed to keep atmospheric concentrations below a high target but the required tax rapidly increases with the stringency of the target. If concentrations are to be kept below 450 ppm CO_2eq, the global carbon tax should reach some \$700/tC in 2015 or so – ten times the recent price of permits in the Emissions Trading System which covers about half of emissions in Europe. Such a carbon tax would roughly double the price of energy in Europe. A 450 ppm CO_2eq concentration would give a 50/50 chance of meeting the declared goal of the European Union and the United Nations to keep global warming below 2°C. However, less ambitious targets would require far lower carbon taxes, and would hardly affect economic growth.

The above discussion about the impacts of climate change suggests that a modest carbon tax can be justified, but that more ambitious goals may be hard to defend.

14.3 Complications

I argue above that climate change is a relatively small problem that can easily be solved. A casual observer of climate policy and the media would have a different impression. Seven things stand in the way of simple solution.

First, there is a demand for an explanation of the world in terms of Sin and a Final Reckoning This is often referred to as Millenarianism. Although many Europeans are nominally secular, fewer are in practice. The story of climate change is often a religious one: emissions (sin) lead to climate change (eternal doom); we must reduce our emissions (atone for our sins). This sentiment is perhaps stronger in Germanic cultures. It has led to an environmental movement (a priesthood) that thrives on preaching climate alarmism, often separated from its factual basis. In order to maximize their membership and income, environmental NGOs meet the demand for scaremongering and moral superiority.

Second, climate policy is perfect for politicians. Climate change is a problem that spans centuries. Substantial emission reduction requires decades and global cooperation. A politician can thus make grand promises about saving the world while shifting the burden of actually doing something (and hurting constituents) to her successor and blaming some foreigner for current inaction.

Third, climate policy allows bureaucrats to create new bureaucracies. Climate policy has been a political priority for about two decades. Emissions have hardly budged, but a vast number of civil servants and larger numbers of consultants and do-gooders have occupied themselves with creating a bureaucratic fiction that something is happening.

Fourth, besides expanded bureaucracies, climate policy can be used to create rents in the form of subsidies, grandparented emission permits, mandated markets and tax breaks. Climate policy thus serves the interests of rent seekers, as well as the interests of policy makers who use rent creation to reward allies.

Fifth, climate policy requires government intervention at the global scale. This antagonizes many, and feeds the fears of right-wing conspiracy theorists. This had led to a movement that attacks climate policy at any opportunity, and extends those attacks to the climate science that underpins that policy, and the scientists who conduct the research. Alarmists have retaliated in kind. The result is polarization, which hampers reasoned discussion on climate policy.

Sixth, greenhouse gas emission reduction is a global public good. The costs of emission abatement are borne by the country that reduces the emissions. The benefits of emission reduction are shared by all of humankind. It is thus individually rational to do very little, and hope that others will do a lot. As every country reasons the same way, nothing much happens. There is no solution to this short of installing a world government.

Seventh, global climate policy has been used as a tactical argument by those who desire a world government for other reasons. Because climate change is such a prominent issue, champions of other worthy causes too have joined the bandwagon. The ultimate goal of climate policy – decarbonisation of the economy – is thus obscured.

14.4 The solution

Any solution to the climate problem should start with acknowledging that we live in a world of many countries, the majority of which jealously guards their sovereignty. That means that climate policy should serve a domestic constituency. Opinion polls in democratic countries have consistently shown over a period of 20 years that a majority is in favour of greenhouse gas emission reduction, even if that means more expensive energy.

Unilateral climate policy is expensive, however. If a country raises its price of energy, but its trading partners do not, business will shift abroad. A country will be more ambitious if it is confident that its neighbours will adopt roughly the same climate policy. The United Nations Framework Convention on Climate Change (UNFCCC) foresees an annual meeting at which countries can indeed pledge their near-term abatement plans and review other countries' progress against previous pledges. This is facilitated by internationally agreed standards on emissions monitoring and reporting. As the actions of trading partners matter most, regional trade organizations, such as the EU, NAFTA, MERCUSOR and ASEAN, should play a bigger role in this process.

The costs of emission reduction vary greatly. It therefore makes sense if countries were allowed to reduce emissions by investing in abatement in other countries. The Kyoto Protocol of the UNFCCC establishes exactly this. Unlike the emissions targets of the Kyoto Protocol, its flexibility mechanisms do not expire.

Therefore, three of the crucial ingredients to a successful climate policy are already in place.

Carbon dioxide is the main anthropogenic greenhouse gas. Fossil fuel combustion is the main source of carbon dioxide emissions. The world would not warm by much if we burn all reserves of conventional oil and gas, the mainstays of the current energy system. Substantial warming requires that we burn considerable amounts of unconventional oil and gas, or use more coal, also in unconventional ways.

Fossil fuel reserves are finite, and the end of conventional oil and gas is in sight. See Figure 2.5. The future energy sector will look radically different from today. The revolution in energy has already begun in the form of tar and shale. Instead of riding the waves of the ongoing revolution, climate policy has focused on creating another energy revolution, hitherto without success. Instead, climate policy should seek to harness the forces of creative destruction that are sweeping the energy sector.

 FURTHER READING

There are many books on climate policy. Good ones include Dieter Helm's *The Carbon Crunch: How We're Getting Climate Change Wrong and How To Fix It* (2012), Nigel Lawson's *An Appeal to Reason: A Cool Look at Global Warming* (2008), William Nordhaus' *The Climate Casino* (2013), and Roger Pielke's *The Climate Fix* (2010). Nick Stern's *Review of the Economics of Climate Change* (2007) is influential but not that good.

Index